AF795922

Conference Proceedings of the Society for Experimental Mechanics Series

Series Editor

Kristin B. Zimmerman, Ph. D.
Society for Experimental Mechanics, Inc.,
Bethel, CT, USA

More information about this series at http://www.springer.com/series/8922

Junlan Wang • Bonnie Antoun • Eric Brown • Weinong Chen
Ioannis Chasiotis • Emily Huskins-Retzlaff • Sharlotte Kramer
Piyush R. Thakre
Editors

Mechanics of Additive and Advanced Manufacturing, Volume 9

Proceedings of the 2017 Annual Conference on Experimental and Applied Mechanics

Springer

Editors
Junlan Wang
Mechanical Engineering
University of Washington
Seattle, WA, USA

Bonnie Antoun
Sandia National Laboratories
Livermore, CA, USA

Eric Brown
Los Alamos National Laboratory
Los Alamos, NM, USA

Weinong Chen
Neil Armstrong Hall
Purdue University
West Lafayette, IN, USA

Ioannis Chasiotis
University of Illinois
Urbana-Champaign, IL, USA

Emily Huskins-Retzlaff
United States Naval Academy
Annapolis, MD, USA

Sharlotte Kramer
Sandia National Laboratories
Albuquerque, NM, USA

Piyush R. Thakre
Dow Chemical Company
Lake Jackson, TX, USA

ISSN 2191-5644 ISSN 2191-5652 (electronic)
Conference Proceedings of the Society for Experimental Mechanics Series
ISBN 978-3-319-62833-2 ISBN 978-3-319-62834-9 (eBook)
DOI 10.1007/978-3-319-62834-9

Library of Congress Control Number: 2017953418

© The Society for Experimental Mechanics, Inc. 2018
This work is subject to copyright. All rights are reserved by the Publisher, whether the whole or part of the material is concerned, specifically the rights of translation, reprinting, reuse of illustrations, recitation, broadcasting, reproduction on microfilms or in any other physical way, and transmission or information storage and retrieval, electronic adaptation, computer software, or by similar or dissimilar methodology now known or hereafter developed.
The use of general descriptive names, registered names, trademarks, service marks, etc. in this publication does not imply, even in the absence of a specific statement, that such names are exempt from the relevant protective laws and regulations and therefore free for general use.
The publisher, the authors and the editors are safe to assume that the advice and information in this book are believed to be true and accurate at the date of publication. Neither the publisher nor the authors or the editors give a warranty, express or implied, with respect to the material contained herein or for any errors or omissions that may have been made. The publisher remains neutral with regard to jurisdictional claims in published maps and institutional affiliations.

Printed on acid-free paper

This Springer imprint is published by Springer Nature
The registered company is Springer International Publishing AG
The registered company address is: Gewerbestrasse 11, 6330 Cham, Switzerland

Preface

Mechanics of Additive and Advanced Manufacturing represents one of nine volumes of technical papers presented at the 2017 SEM Annual Conference and Exposition on Experimental and Applied Mechanics organized by the Society for Experimental Mechanics and held in Indianapolis, IN, June 12–15, 2017. The complete Proceedings also includes volumes on: *Dynamic Behavior of Materials*; *Challenges In Mechanics of Time-Dependent Materials*; *Advancement of Optical Methods in Experimental Mechanics*; *Mechanics of Biological Systems, Materials and other topics in Experimental and Applied Mechanics*; *Micro- and Nanomechanics*; *Mechanics of Composite, Hybrid and Multifunctional Materials*; *Fracture, Fatigue, Failure and Damage Evolution*; and *Residual Stress, Thermomechanics and Infrared Imaging, Hybrid Techniques and Inverse Problems*.

Mechanics of additive and advanced manufacturing is an emerging area due to the unprecedented design and manufacturing possibilities offered by new and evolving advanced manufacturing processes and the rich mechanics issues that emerge. Technical interest within the society spans several other SEM technical divisions such as: composites, hybrids and multifunctional materials, dynamic behavior of materials, fracture and fatigue, residual stress, time-dependent materials, and the research committee.

In this inaugural track in SEM 2017, the topic of mechanics of additive and advanced manufacturing included in this volume covers design, optimization, experiments, computations, and materials for advanced manufacturing processes (3D printing, micro- and nano-manufacturing, powder bed fusion, directed energy deposition, etc.) with particular focus on mechanics aspects (e.g., mechanical properties, residual stress, deformation, failure, rate-dependent mechanical behavior, etc.).

Seattle, WA, USA	Junlan Wang
Livermore, CA, USA	Bonnie Antoun
Los Alamos, NM, USA	Eric Brown
West Lafayette, IN, USA	Weinong Chen
Urbana-Champaign, IL, USA	Ioannis Chasiotis
Annapolis, MD, USA	Emily Huskins-Retzlaff
Albuquerque, NM, USA	Sharlotte Kramer
Lake Jackson, TX, USA	Piyush R. Thakre

Contents

Chapter 1
Fracture Properties of Additively Manufactured Acrylonitrile-Butadiene-Styrene Materials

Kevin R. Hart and Eric D. Wetzel

Abstract Additively Manufactured (AM) parts exhibit orthotropic behavior when loaded as a result of the layer-by-layer assembly commonly utilized. While previous authors have studied the effect of layer orientation on the tensile, flexural, and impact response of AM parts, the effect of layer orientation on the fracture response is not well established. Here we explore the effect of layer orientation on the fracture properties of Acrylonitrile-Butadiene-Styrene (ABS) materials fabricated through the Fused Filament Fabrication (FFF) process. Critical fracture toughness values of Single Edge Notch Bend (SENB) specimens with a pre-crack oriented either parallel or perpendicular to the direction of layer-by-layer assembly were compared. Results show that the inter-laminar fracture toughness (fracture between layers) is approximately one order of magnitude lower than the cross-laminar toughness (fracture through layers) of similarly manufactured parts. Contrasting brittle and ductile fracture behavior is observed for inter-laminar and cross-laminar crack propagation, respectively, demonstrating that the elastic-plastic response of AM ABS parts is governed by the orientation of the layers with respect to the direction of crack propagation.

Keywords Additive manufacturing • J-integral • Fracture mechanics • Polymer • Single edge notch bend

SENB specimens measuring 100 $\times$ 20 $\times$ 10 mm were fabricated using a Taz 6 FFF desktop printer (Lulzbot; Loveland, Co), with ABS M30 filament (Stratasys; Eden Prarie, MN). Printed specimens are described as laminates, where each laminae is a layer of material constructed parallel to the print bed with a nominal thickness equal to the layer height (0.22 mm). Samples with variable crack-tip/laminae orientation angles, θ, were fabricated. Specimens for inter-laminar fracture testing were fabricated with the longest part edge oriented perpendicular to the build plate (vertically) and had a crack-tip/lamina orientation of $\theta = 0°$. Specimens for cross-laminar fracture testing were constructed with the longest part edge oriented parallel to the build plate (horizontally) and had a crack-tip/lamina orientation of $\theta = 90°$. Specimens to investigate crack kinking behavior were constructed with the longest edge oriented at an oblique angle of 75° relative to the vertical (oblique). Orientations of vertically, horizontally, and obliquely printed SENB samples are depicted in Fig. 1.1a.

Three-point SENB testing was performed on a benchtop load frame. Fracture response of the specimens varied depending on the crack-tip/lamina orientation angle, θ. Vertically printed samples with $\theta = 0°$ displayed inter-laminar fracture and had a load/displacement response consistent with brittle behavior (Fig. 1.2a). Horizontally printed samples with $\theta = 90°$ displayed cross-laminar fracture and had a load/displacement response consistent with ductile behavior (Fig. 1.2a). Obliquely printed samples with $\theta = 75°$ displayed mixed cross- and inter-laminar fracture, resulting in mixed ductile and brittle crack propagation (Fig. 1.2a). Since fracture behavior was inconsistent across sample types, differing methods of fracture analysis were required to determine the samples' fracture toughness. Analysis methods are reserved for further reading in Hart et al. [1]. Results of the fracture toughness analysis reveal that the critical elastic-plastic strain energy release rates for inter-laminar and cross-laminar fracture were 256 $\pm$ 84 J/m^2 and 2260 J/m^2, respectively, demonstrating that significantly more energy is required to propagate a crack across the laminae than between them.

This study confirms that the low strength of weld-lines between or among laminae, relative to within a deposited polymer trace, limits the toughness of the overall part. However, the likelihood of catastrophic failure due to fracture in an AM structure can be reduced by considering anticipated stress fields when designing laminae orientations. Naturally orthotropic materials such as wood, bone, and nacre, have all adapted to capitalize on the increased toughness afforded by flaw/laminae alignment to impart global toughness in spite of constituents which may be individually brittle [2–7]. We believe that there exists considerable opportunity to engineer higher performance AM parts by exploiting these design principles to create tailored fracture and failure behaviors.

K.R. Hart (✉) • E.D. Wetzel
Weapons and Materials Research Directorate, US Army Research Laboratory, 4600 Deer Creek Loop, Aberdeen Proving Ground, Aberdeen, MD 21005, USA
e-mail: kevin.hart10.ctr@mail.mil

© The Society for Experimental Mechanics, Inc. 2018

J. Wang et al. (eds.), *Mechanics of Additive and Advanced Manufacturing, Volume 9*,
Conference Proceedings of the Society for Experimental Mechanics Series, DOI 10.1007/978-3-319-62834-9_1

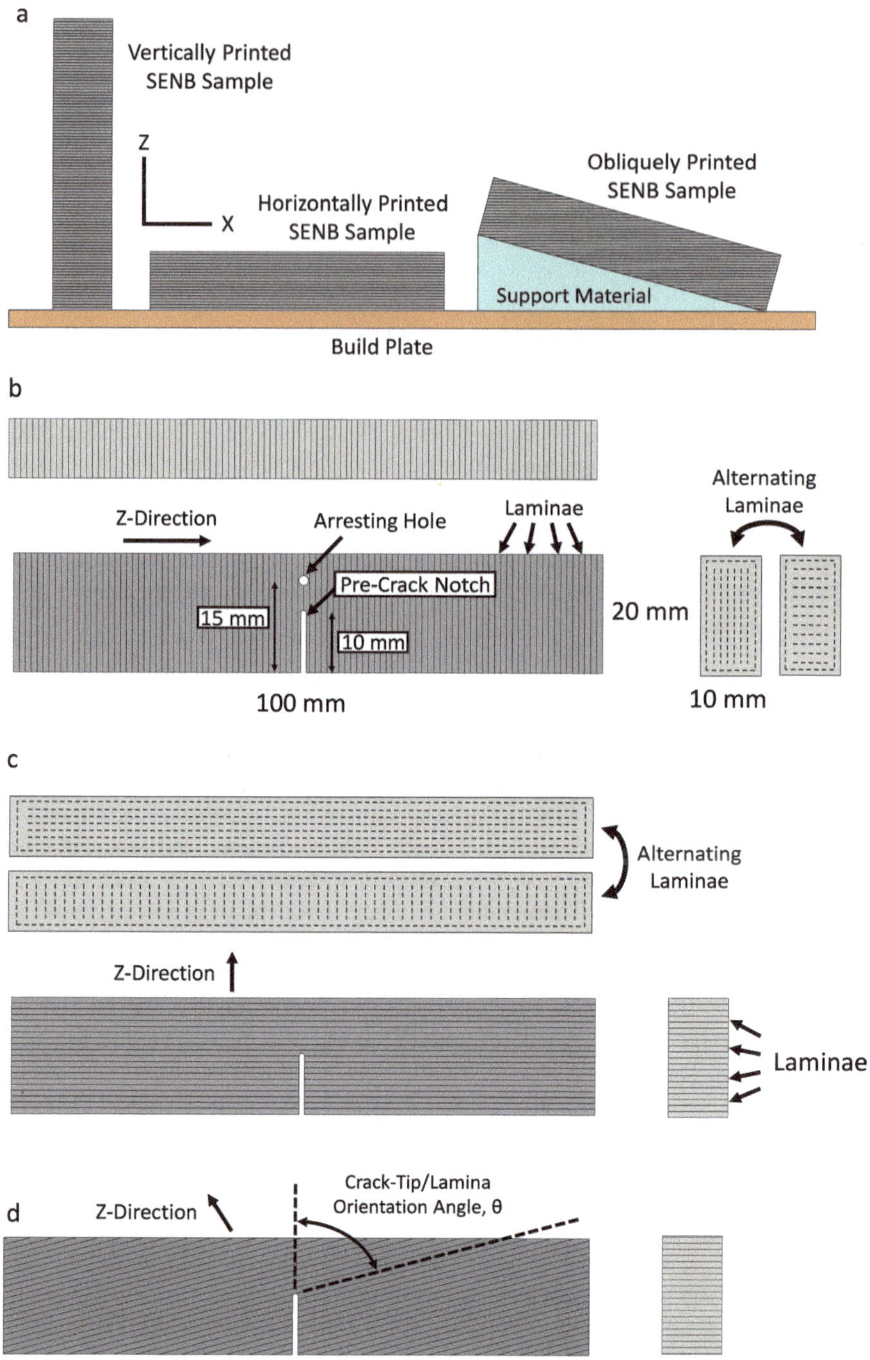

Fig. 1.1 Illustrations of SENB fracture specimens. (**a**) Orientation of vertically, horizontally, and obliquely printed SENB specimens with respect to the build plate. (**b**) Dimensions of a vertically printed SENB sample after removal from the build plate, preparation of the pre-crack, and machining of the arresting hole. Here the laminae run parallel to the pre-crack, facilitating inter-laminar fracture. (**c**) Illustration of a horizontally printed SENB sample after removal from the build plate and preparation of the pre-crack. Here the laminae run perpendicular to the pre-crack, facilitating cross-laminar fracture. (**d**) Illustration of an obliquely printed SENB sample after removal from the build plate and preparation of the pre-crack. Laminae are oriented at an angle of θ with respect to the pre-crack notch. Laminae, pre-crack notches, and raster lines are not to scale. *Dashed lines* indicate position and orientation of weld-lines (not to scale) between deposition traces within each laminae. Image adopted from Hart et al. [1]

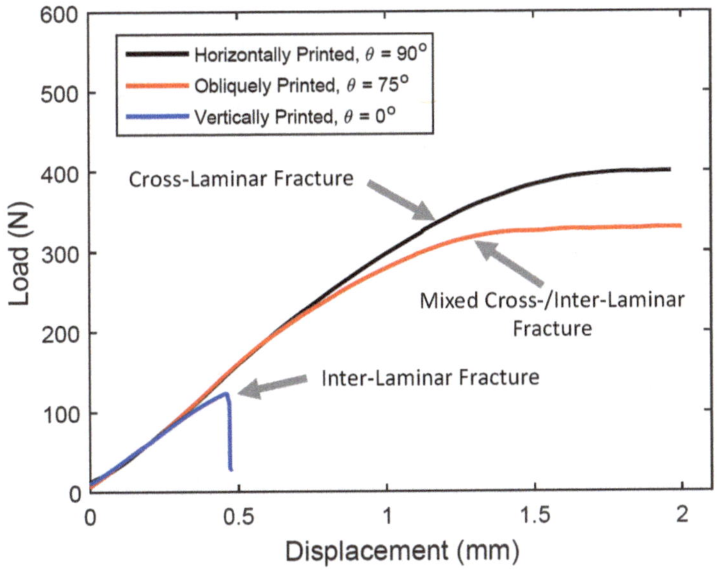

Fig. 1.2 Load vs. displacement curves for SENB specimens with crack-tip/lamina orientation angles of $\theta = 0°$, $\theta = 75°$, and $\theta = 90°$. Vertically printed SENB specimens with $\theta = 0°$ exhibit brittle, inter-laminar fracture. Horizontally printed SENB specimens with $\theta = 90°$ exhibit ductile, cross-laminar fracture. Obliquely printed SENB specimens with $\theta = 75°$ exhibit mixed cross-/inter-laminar fracture behavior. Image adopted from Hart et al. [1]

References

1. Hart, K.R., Wetzel, E.D.: Fracture behavior of additively manufactured acrylonitrile butadiene styrene (ABS) materials. Eng. Fract. Mech. **177**, 1–13 (2017)
2. Bechtle, S., Fett, T., Rizzi, G., Habelitz, S., Schneider, G.A.: Mixed-mode stress intensity factors for kink cracks with finite kink length loaded in tension and bending: application to dentin and enamel. J. Mech. Behav. Biomed. Mater. **3**, 303–312 (2010)
3. Yan, J., Mecholsky Jr., J.J., Clifton, K.B.: How tough is bone? Application of elastic-plastic fracture mechanics to bone. Bone. **40**, 479–484 (2007)
4. Jernkvist, L.O.: Fracture of wood under mixed mode loading II. Experimental investigation of Picea abies. Eng. Fract. Mech. **68**(November 2000), 565–576 (2000)
5. Jernkvist, L.O.: Fracture of wood under mixed mode loading I. Derivation of fracture criteria. Eng. Fract. Mech. **68**(November 2000), 549–563 (2001)
6. Wang, R., Gupta, H.S.: Deformation and fracture mechanisms of bone and nacre. Annu. Rev. Mater. Res. **41**, 41–73 (2011)
7. Zimmermann, E.A., Launey, M.E., Barth, H.D., Ritchie, R.O.: Mixed-mode fracture of human cortical bone. Biomaterials. **30**, 5877–5884 (2009)

Chapter 2
Complex Modulus Variation by Manipulation of Mechanical Test Method and Print Direction

Megan L. Liu, Katherine K. Reichl, and Daniel J. Inman

Abstract 3D printing technologies have made creating prototypes with complex geometries relatively simple thus it has become an increasingly popular method for creating prototypes in a research setting. Therefore, it is crucial to understand the properties of the materials being used. This paper examines the effects of printing direction and testing method type on the complex modulus of viscoelastic materials printed using the Objet Connex 3D Printer from Stratasys. Because of its ability to print multiple materials in a single print job, this printer is a popular choice to create models. Throughout these tests the sample material will be kept constant to isolate the effects of print direction and test performed. DM 8430 is produced by mixing VeroWhitePlus™ and TangoPlus™ in a specific ratio. Since the 3D printer threads and smooths the sample uniaxially, the print direction of the sample can be manipulated by changing the orientation at which the sample is placed on the printer. Two different print directions, that are perpendicular with respect to each other, will be examined. The two test methods that will be used to determine the complex modulus are the Dynamic Mechanical Analysis (DMA) test, which examines the tensile behavior of the material, and the vibrating beam test, which examines the bending behavior. The goal is to gain greater insight into the uncertainty in the complex modulus that results from changing the test and printing direction used to determine this value. This will be done by performing a total of four tests. For each testing method, DMA and vibrating beam, the complex modulus will be found for two samples of different print direction, vertical and horizontal. These results will permit a greater understanding of the amount of variability produced by print direction.

Keywords Additive manufacturing • Complex modulus • Viscoelastic • Passive damping • Dynamic modulus analysis

2.1 Introduction

Additive manufacturing has become increasing popular across almost all engineering fields. The field of interest to the authors is in applications for metastructures to be used for vibration suppression [1, 2]. The specific printer of interest is the Objet Connex printer which is capable of printing rubber-like materials. These materials exhibit viscoelastic properties which is of particular interest for vibration suppression [3]. The purpose of this paper is towards the determination of the viscoelastic properties of the printed materials.

The Object Connex printer uses PolyJet printing technology which works like an inkjet printer. The parts are made by depositing many small dots of material and then curing the resin resulting in an end material that appears homogenous. Because of the digital natural of this method, these materials are referred to as digital materials (DM). This method allows the printer to easily mix two different base materials in various ratios to create a gradient of material with various hardness levels [4]. This method also allows for parts made in a single print with both rigid and viscoelastic materials.

Using this technique and the many base materials available, the Object Connex printer has the capability to create many materials but for this paper the focus will be on the digital materials created using the two base materials, VeroWhitePlus™ and TangoPlus™. VeroWhitePlus™ is a rigid opaque material and TangoPlus™ is a rubber-like transparent material [5]. Using these two base materials, ten different digital materials can be created [6]. The material studied here is DM 8430 which is the third stiffest material. In the field of active composites and origami, H. Jerry Qi has done extensive work using the Objet Connex 3D printer and his papers provide many details the actual mechanisms the 3D printer utilizes [7].

M.L. Liu
Department of Material Science and Engineering, University of Michigan, François-Xavier Bagnoud Building, 1320 Beal Avenue, Ann Arbor, MI 48109, USA

K.K. Reichl (✉) • D.J. Inman
Department of Aerospace Engineering, University of Michigan, François-Xavier Bagnoud Building, 1320 Beal Avenue, Ann Arbor, MI 48109, USA
e-mail: reichl@umich.edu

© The Society for Experimental Mechanics, Inc. 2018

J. Wang et al. (eds.), *Mechanics of Additive and Advanced Manufacturing, Volume 9*,
Conference Proceedings of the Society for Experimental Mechanics Series, DOI 10.1007/978-3-319-62834-9_2

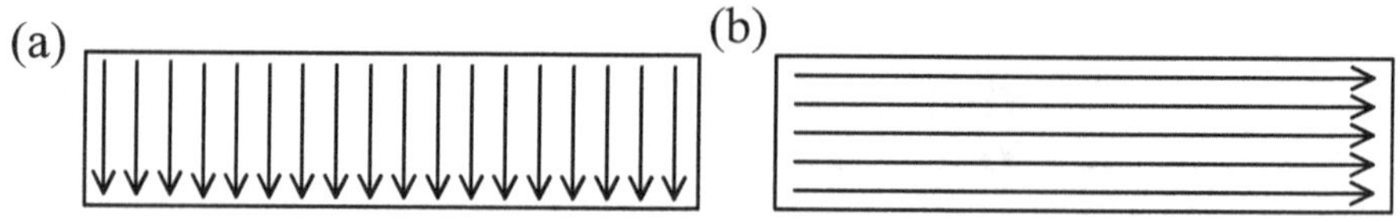

Fig. 2.1 Schematic of print directions utilized (**a**) horizontal direction (**b**) vertical direction

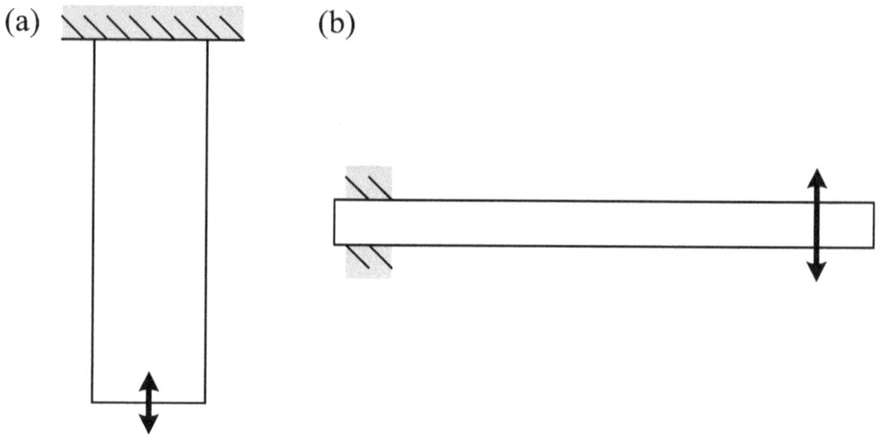

Fig. 2.2 Schematic of testing configurations utilized (**a**) tensile configuration (**b**) single cantilever configuration

In this paper, two different print directions are studied in addition to two different testing configurations. The two print directions are depicted in Fig. 2.1 where the arrows represent the direction in which the material was printed. The Object Connex always prints parallel rows of materials. The reasoning behind these two testing methods is to determine how these the print direction affects the dynamic modulus properties. The print direction creates visible differences in the parts.

Two different testing configurations were utilized to determine the effects of the testing configuration. One sample was tested in tension and the other was tested using the single cantilever orientation (Fig. 2.2). The storage modulus, loss modulus, and loss factor values obtained from these tests are compared and the differences that result from a change in testing method and print direction are analyzed.

2.2 Viscoelastic Material Behavior

The viscoelastic phenomen can be seen in the dynamic response of a viscoelastic material due to a sinusoidal load. For an elastic material if a harmonically varying stress is applied to the material, the strain response will be at that same frequency and it phase with the load. For a viscoelastic material this is not the case. The response is still at the same frequency as the input but has a phase lag. If a time-varying stress $\sigma(t)$ is applied to a viscoelastic material as follows

$$\begin{aligned} \sigma(t) &= \sigma_0 \sin(\omega t) \\ \varepsilon(t) &= \varepsilon_0 \sin(\omega t - \delta) \end{aligned} \tag{2.1}$$

where ω is the frequency of the forcing function and strain-response of the material is $\varepsilon(t)$. The phase lag is denoted as δ. From these relatonships, a dynamic stiffness (or complex stiffness) can be expressed as a complex number.

$$\frac{\sigma(t)}{\varepsilon(t)} = E'(\omega) + i\ E''(\omega) = E(\omega)[1 + i\,\mu(\omega)] \tag{2.2}$$

where $E^{'}$ is called the storage modulus and $E^{''}$ is known as the loss modulus. The modulus can also be expressed by factoring out the storage modulus which leaves the loss factor, μ. The modulus values vary with both frequency and temperature [8].

Using the time-frequency equivalence, we can assume that the complex modulus value at a frequency, f_1 and a temperature T_1 is equal to the value at any other frequency f_2 and some temperature T_2 such that the following relationship can be made

$$E(f_1,\ T_1) = E\ (f_2\alpha(T_2)) \tag{2.3}$$

where $\alpha(T_2)$ is called the shift factor and describes the relationship between frequency and temperature. Using the shift factor, the effects of both temperature and frequency can be combined into a single variable known as the reduced frequency, $f_r(f,\ T) = f\alpha(T)$. The shift factor relationship is determined by testing a material at multiple frequencies and temperatures. For each temperature, a shift factor value is determined graphically by examining the modulus versus reduced frequency plots. Varying the shift factor causes the data at that temperature to shift on the reduced frequency scale. Once the shift factors for each temperature are determined, the $\log[\alpha(T)]$ is plotted versus $1/T$. This relationship resembles a slightly curved line, which here is approximated as straight line and the Arrhenius shift factor equation is fit to the data. The Arrhenius shift factor equation takes on the following form

$$\log[\alpha(T)] = T_A\left(\frac{1}{T} - \frac{1}{T_0}\right) \tag{2.4}$$

where T_0 is any arbitrarily selected reference temperature and T_A is the slope of the line and is related to the activation energy. Using this relationship, the complex modulus and loss factor can be expressed as follows

$$E(f,T) = E\left[f\ \exp\left({}^{-2.303T_A}/_T\right)\exp\left({}^{2.303T_A}/_{T_0}\right)\right] \quad \eta(f,T) = \eta\left[f\ \exp\left({}^{-2.303T_A}/_T\right)\exp\left({}^{2.303T_A}/_{T_0}\right)\right] \tag{2.5}$$

2.3 Testing Methods

The complex modulus of the DM 8430 material was obtained using the Dynamic Mechanical Analysis (DMA) Q800 machine by Thermal Analysis using the tensile and single cantilever configurations as explained above. Two samples were tested for each configuration each which used a different print direction.

For the tensile configuration, the testing procedure followed the ASTM D5026 standard [9]. Figure 2.3 shows a setup of the of test. The sample specimen was held in place by two clamps, the upper of which was fixed while the bottom moved cyclically. Both samples were tested at 9 temperatures varying linearly between 25 and 105 °C. A soak time of 5 min was used at each temperature to ensure the sample reached the isothermal state. At each temperature, the sample was tested at 13 frequencies varying logarithmically between 100 and 0.10 Hz. The dimensions of the samples were approximately 9.5 × 1 × 5 mm.

Using the cantilevered configuration, the sample was tested in flexure using the Q800 single cantilever fixture. This procedure closely follows the ASTM D5418 standard. The standard presented pertains to a dual cantilever beam and here a single cantilever was used [10]. A fixed clamp held one side of the beam, and the other moved cyclically at the specified frequency of the test sequence. These samples were tested at 9 temperatures varying linearly between 35 and 90 °C once

Fig. 2.3 DMA tension test configuration

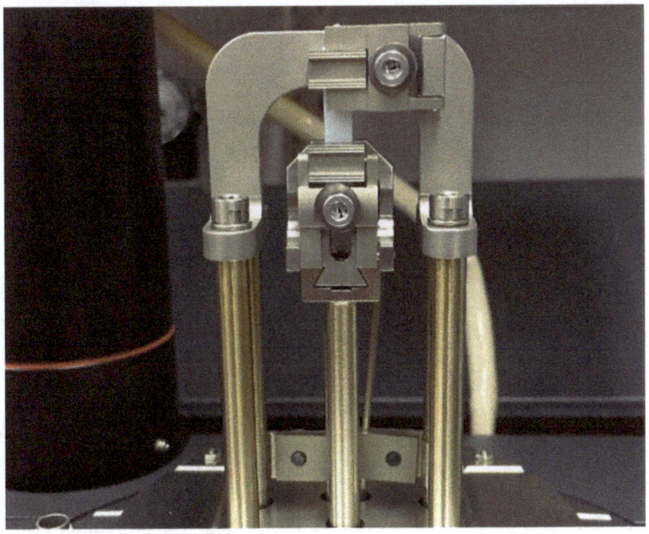

again with a soak time of 5 min. At each temperature, the specimen was tested at 10 frequencies varying logarithmically between 18 and 0.10 Hz. The cantilevered specimens had dimension of 35 × 3 × 12 mm.

The DMA Q800 software collects and analyzes the data for each temperature and frequency. Based on the geometry of the testing specimen, the software outputs the storage modulus and the loss factor for the material at the specified temperatures and frequencies. Next, the data is manually analyzed further to determine the relationship between temperature and frequency. The shifting procedure described above is used to determine the relationship between the shift factors, α and the temperature T. This allows the complex modulus data to be plotted against a single variable, reduced frequency. For all tests a reference temperature of 65° C was used. The parameters for the best fit line for the Arrhenius fit is also reported.

2.4 Results

The final shifted results from all four of the configurations tested are presented in this section these configurations are summarized in Table 2.1. The results for the tensile configuration are shown in Figs. 2.4 and 2.5 and for the cantilevered configuration are shown in Figs. 2.6 and 2.7. The complex modulus values are plotted versus reduced frequency and the logarithmic value of shift factor used to determine the reduced frequency are plotted versus the inverse of the temperature along with the Arrhenius fit. The values of the Arrhenius fit line are reported in Table 2.1.

From these plots, the horizontal and the vertical printing directions are virtually indistinguishable from each other. The results lead us to believe that the in-place print direction does not affect the complex modulus values. Next, a comparison of the testing configurations is shown in Fig. 2.8. The testing configuration clearly makes a difference in the complex modulus values of the material. This difference is much more noticeable in the storage modulus values than the loss factor data. In the loss factor data, the peak in the tensile configuration data has a higher value than that of the cantilever configuration.

Table 2.1 Arrhenius shift factor equation fit

Configuration	Printing direction	Reference temperature [°C]	Arrhenius temperature [K]
Tensile	Horizontal	65	12,372
Tensile	Vertical	65	12,804
Cantilever	Horizontal	65	14,472
Cantilever	Vertical	65	15,667

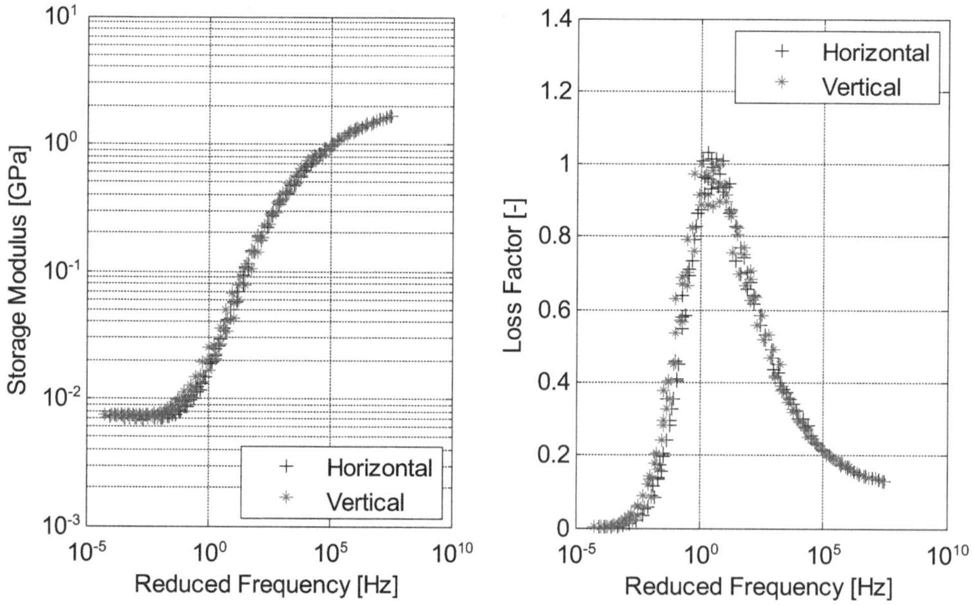

Fig. 2.4 Complex modulus plots for the vertical and horizontal printing directions of the tensile configuration with a reference temperature of 65 °C

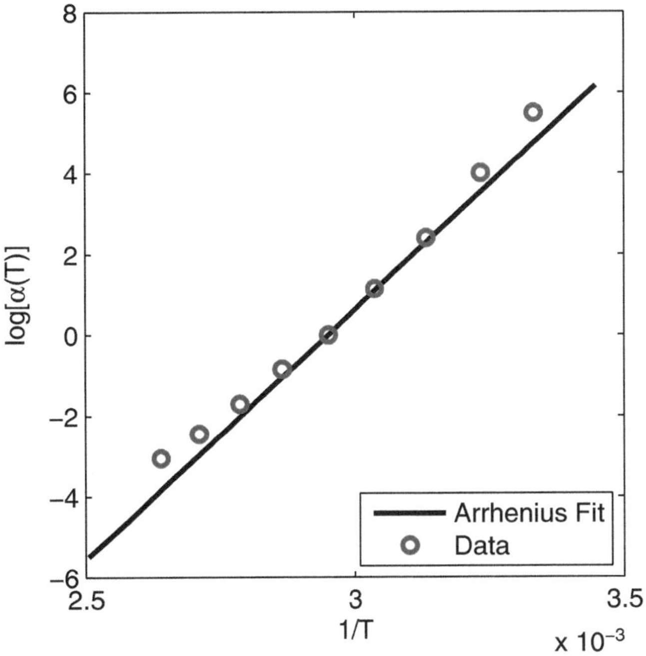

Fig. 2.5 Shift factor relationship and Arrhenius fit for the tensile configuration

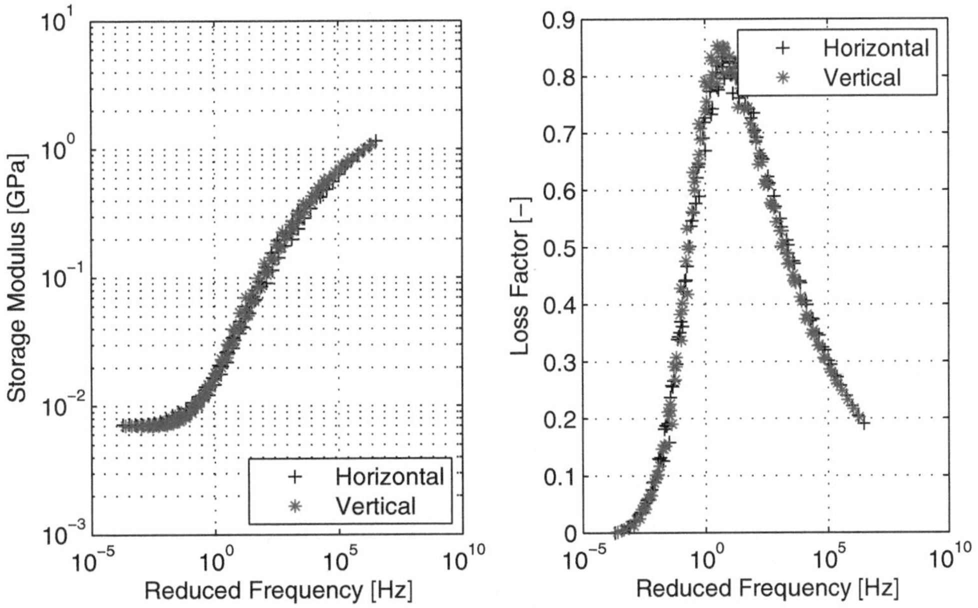

Fig. 2.6 Complex modulus plots for the vertical and horizontal printing directions of the cantilevered configuration with a reference temperature of 65 °C

2.5 Conclusions

Firstly, this work shows that the DM 8430 material exhibits significant viscoelastic material behavior which should be taken into consideration when modelling this material. This is particularly important when temperature change is occurring. In addition, this work provides valuable insight into the effects of print direction on the modulus properties of the material. The results show that variations in the in-plane print direction does not affect the response of the material. This was demonstrated in both the tensile and cantilevered configurations. The testing configuration does affect the complex modulus values, thus when using this data for modeling the configuration that most closely resembles the structure being modeled should be used.

Fig. 2.7 Shift factor relationship and Arrhenius fit for the cantilevered configuration

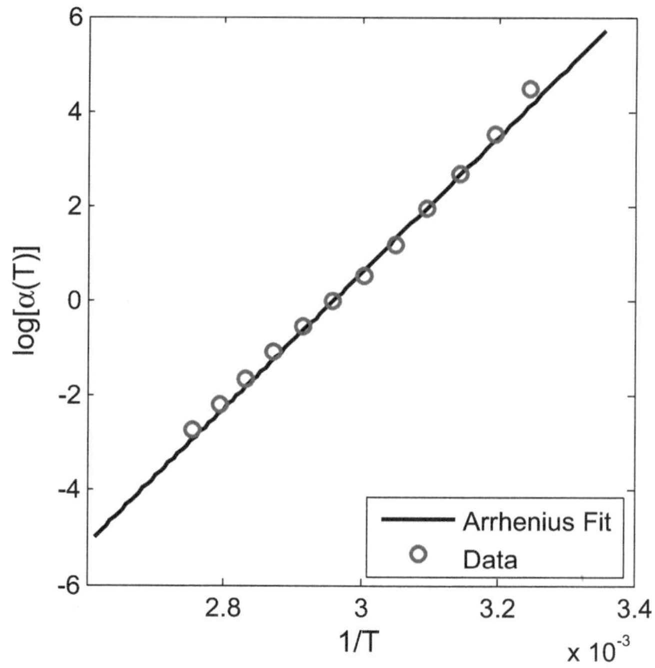

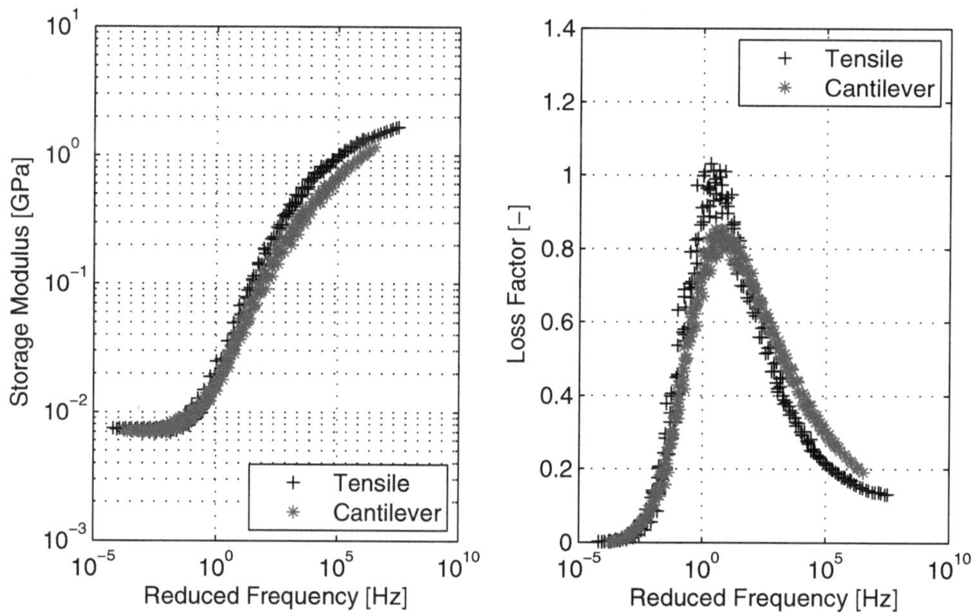

Fig. 2.8 Complex modulus data comparing the tensile and cantilevered configurations with a reference temperature of 65 °C

Acknowledgements This work is supported in part by the US Air Force Office of Scientific Research under the grant number FA9550-14-1-0246 "Electronic Damping in Multifunctional Material Systems" monitored by Dr. BL Lee and in part by the University of Michigan.

References

1. Reichl, K. K., Inman, D. J.: Modelling of low-frequency broadband vibration mitigation for a bar experiencing longitudinal vibrations using distributed vibration absorbers. In: 20th International Conference on Composite Materials, Copenhagen, 2015
2. Hobeck, J. D., Laurant, C. M. V., Inman, D. J.: 3D printing of metastructures for passive broadband vibration suppression. In: 20th International Conference on Composite Materials, Copenhagen, 2015
3. Ge, Q., Dunn, C.K., Qi, H.J., Dunn, M.L.: Active origami by 4D printing. Smart Mater. Struct. **23**(94007), 1–15 (2014)

4. Objet Connex 3D Printers [Online]. Available: http://www.stratasys.com/3d-printers/design-series/connex-systems (2015). Accessed 17 Oct 2015
5. Stratasys PolyJet Materials Data Sheet [Online]. Available: http://usglobalimages.stratasys.com/Main/Secure/Material Specs MS/PolyJet-Material-Specs/Digital_Materials_Datasheet.pdf?v=635581278098921962 (2014). Accessed 17 Oct 2015
6. Stratasys Digital Materials Data Sheet [Online]. Available: http://usglobalimages.stratasys.com/Main/Secure/Material Specs MS/PolyJet-Material-Specs/Digital_Materials_Datasheet.pdf?v=635581278098921962 (2015). Accessed 17 Oct 2015
7. Ge, Q., Mao, Y., Yu, K., Dunn, M. L., Qi, H. J.: Active composites and 4D printing. In: 20th International Conference on Composite Materials, Copenhagen, 2015
8. Lakes, R.S.: Dynamic behavior. In: Viscoelastic Materials, pp. 55–90. Cambridge University Press, New York (2009)
9. ASTM: Standard Test Method for Measuring the Plastics: Dynamic Mechanical Properties of Plastics in Tension Properties: In Tension 1, no. D5026 – 15, pp. 1–5. (2001)
10. ASTM: Standard Test Method for Plastics: Dynamic Mechanical Properties: In Flexure (Three-Point Bending), no. D5418 – 15, pp. 1–4. (2007)

Chapter 3
A New Heat Transfer Simulation Model for Selective Laser Melting to Estimate the Geometry of Cross Section of Melt Pool

Hong-Chuong Tran and Yu-Lung Lo

Abstract For simulating the geometry of melt pool's cross section more accurately in Selective Laser Melting (SLM) process, modeling the volumetric heat source is considered as the key issue. In the present study, a three dimensional finite element heat transfer simulation with new volumetric heat source is proposed to simulate the geometry of melt pool's cross section during SLM process. The approach to build up the volumetric heat source is based on the modified sequential addition method to simulate the powder bed and the Monte Carlo ray tracing simulation in Zemax to calculate the energy distribution along the depth of metal powder layer. For validating the proposed simulation model, the simulated result of the contact-width between melt pool and substrate is compared with experimental result presented in the existing literature.

Keywords Ray tracing • Powder bed • Absorption • Volumetric heat source • Melt pool

3.1 Introduction

Selective Laser Melting (SLM) is a typical Additive Manufacturing (AM) process for manufacturing 3D parts by selectively melting specific areas of metal powder layers using thermal energy of a controlled laser beam. Processing parameters of SLM can be classified into four types: laser related parameters, scan related parameters, powder related parameters, temperature-related parameters [1]. These parameters are mutual interacting during manufacturing process and create profound impact on the mechanical properties of final parts. In addition, relying on experimental trial and error for finding the optimal processing parameters in SLM is expensive and time consuming. Consequently, simulation model can be considered as an effective tool for the estimation of suitable processing parameters.

In SLM process, as the laser beam irradiates powder layer, it will penetrate through the depth of powder layers due to multiple reflections of laser radiation between powder particles. Therefore, in order to improve the accuracy in modeling the heat source for finite element heat transfer simulation, the authors in literatures [2, 3] proposed methods to construct the volumetric heat source.

In the present study, a new methodology is proposed to build up the volumetric heat source. The proposed approach can be described as follows. Firstly, a modified sequential addition simulation model to simulate the metal powder layer is built up. Then, the geometry of powder bed is exported to Zemax for performing Monte Carlo ray tracing simulation. Then, the absorption profile along the depth of powder bed is obtained. Accordingly, the equation describing the volumetric heat source is proposed. Finally, the built up volumetric heat source is used in finite element heat transfer simulation implementing in COMSOL Multiphysics software and Matlab.

H.-C. Tran
Department of Mechanical Engineering, National Cheng Kung University, Tainan, Taiwan
e-mail: hongchuong3389@gmail.com

Y.-L. Lo (✉)
Department of Mechanical Engineering, National Cheng Kung University, Tainan, Taiwan

Advanced Optoelectronic Technology Center, National Cheng Kung University, Tainan, Taiwan
e-mail: loyl@mail.ncku.edu.tw

© The Society for Experimental Mechanics, Inc. 2018

J. Wang et al. (eds.), *Mechanics of Additive and Advanced Manufacturing, Volume 9*,
Conference Proceedings of the Society for Experimental Mechanics Series, DOI 10.1007/978-3-319-62834-9_3

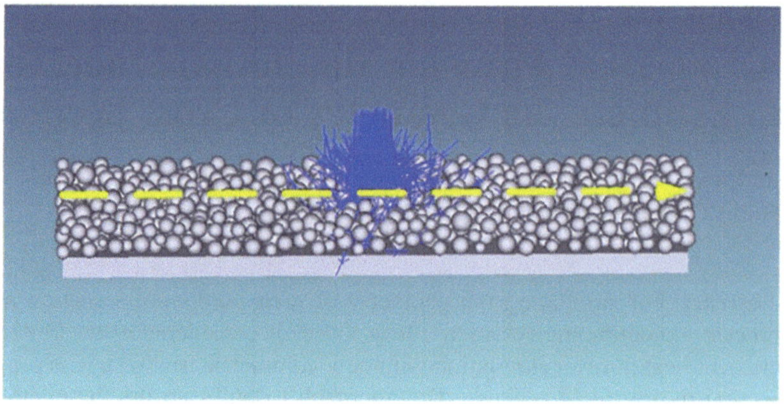

Fig. 3.1 Simulation in Zemax

3.2 Volumetric Heat Source Modeling

A method to build up the volumetric heat source can be described as follows. Firstly, the modified sequential addition simulation model developed in our research group is employed to simulate the powder bed used in SLM. Notably, in literature [3], the mean diameter of powder particles is given as $d_{mean} = 20$ μm but the information of powder size distribution was not specified. However, in the real SLM applications, the powder size distributions usually conforms Gaussian distribution. Therefore, in the present study, the powder layer with Gaussian distribution which has mean diameter equal to 20 μm and the standard deviation of diameter of 2.5 μm is simulated. Secondly, the geometry details of the simulated metal powder layer were imported to Zemax using self-written macro in order to perform Monte Carlo ray tracing simulations. Figure 3.1 demonstrates the ray tracing simulation built up in Zemax in the present study. As shown, illuminating source is set as Gaussian laser beam. Following the ray tracing simulations described above, substantial post scripting written in Matlab was used to extract the ray database of Zemax to calculate the absorbed energy along the depth of metal powder layer deposited on the substrate. This absorption profile along the depth of metal powder layer can be described mathematically as dA/dz where A is the laser absorption of powder layer and z is the coordinate along the depth of powder bed.

After the absorption profile along the depth of powder bed is obtained, the equation for mathematically describing the volumetric heat source is proposed as

$$q(x, y, z, t) = \frac{2P}{\pi r_0^2} \exp\left[-\frac{2\left((x - vt)^2 + y^2\right)}{r_0^2} \right] \frac{dA}{dz} \tag{3.1}$$

where P is the laser power, the unit of P is Watt; r_0 is the radius of laser beam at which laser energy is $I = I_0/e^2$, the unit of r_0 is meter; v is the scanning speed of laser beam, the unit of v is m/s; t is the simulation time, the unit of t is second; $\frac{dA}{dz}$ represents the laser absorption along the depth of powder layer, the unit of $\frac{dA}{dz}$ is 1/μm. The unit of the calculated laser energy density denoted as $q(x, y, z, t)$ is W/m^3.

3.3 Finite Element Modeling

A non-linear transient thermal analysis is built up in COMSOL Multiphysics and Matlab. The dimension of solid substrate is set as 600 × 300 × 150 μm and the dimension of powder region is specified as 600 × 300 × 50 μm. It is noted that the type of mesh size is hexahedron element. By performing trial simulations, it is known that the solution is convergence by using the mesh size of 8 μm. The time step for running the simulation is set as 0.0001 s.

3.4 Results and Discussion

Figure 3.2a shows the temperature distribution on the surface of metal powder layer and substrate at scanning speed $v = 120$ mm/s and laser power P = 45 W. The geometry of melt pool's top view is displayed in Fig. 3.2b. It is noted that the region inside the red contour showed in Fig. 3.2a, b is the melt pool and the temperature in this region is beyond the melting temperature (1648 K). Figure 3.3a displays the temperature distribution at cross section of melt pool. In order to better visualize the shape of melt pool's cross section and how it interacts with substrate, in Fig. 3.3b, the color code is set so that the region having temperature exceeds the melting point is in gray color, and the area that has temperature lower than the melting point is not displayed. Notably, in Fig. 3.3b, the blue region is substrate and red region is powder. Figure 3.3c is the experimental result presented in literature [3], and this experiment was conducted at laser power 45 W and the laser scanning speed 120 mm/s. It is noted that the effect melt flow which is driven by the surface tension force on the formation of simulated melt pool are not are not considered in the present study.

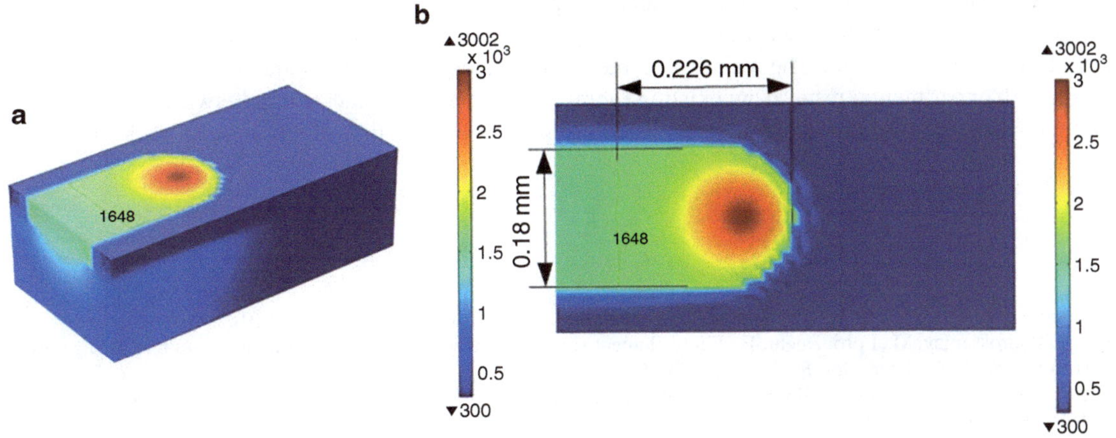

Fig. 3.2 (**a**) Surface temperature distribution; (**b**) Top view of metal powder layer

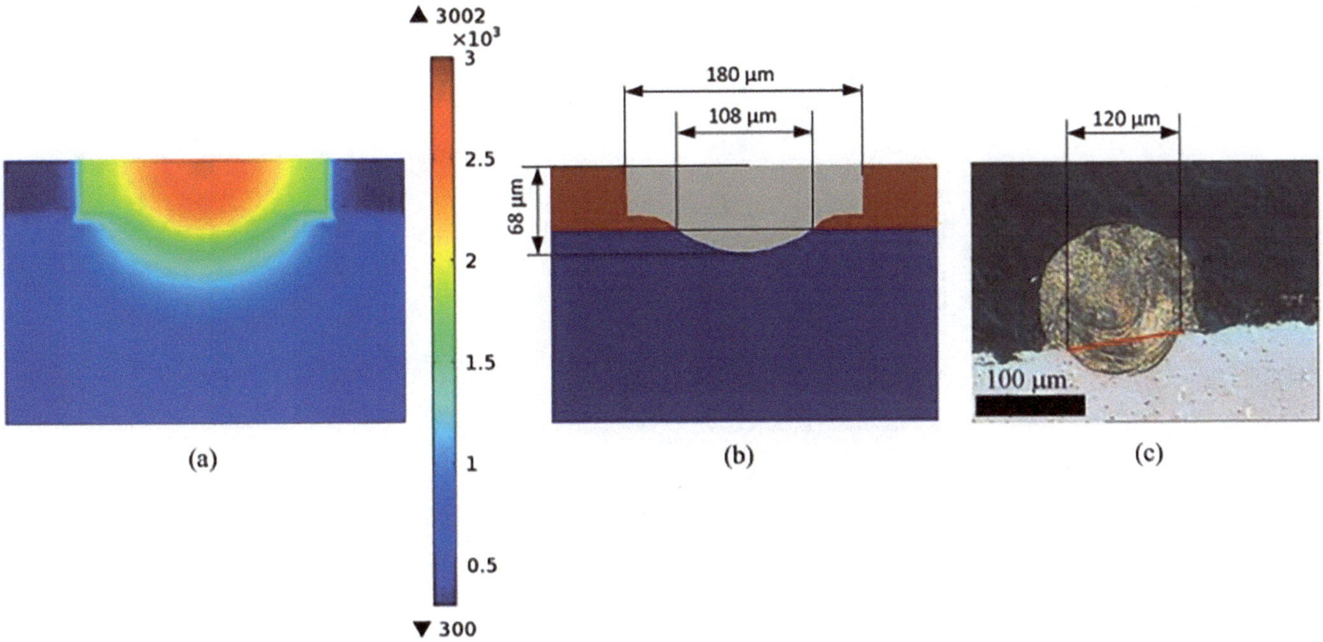

Fig. 3.3 (**a**) Temperature distribution at the cross section of melting pool; (**b**) geometry of the simulated melt pool's cross section; (**c**) Cross section of melted pool from literature [3]

For validating the proposed simulation model, the simulated results of contact-width between melt pool and substrate are compared with experimental results obtained in literature [3]. It is noted that the contact-width between melt pool and substrate obtained from experiment of study [3] is indicated by red color in Fig. 3.3c. From Fig. 3.3b, c, it is observed that the geometry of simulated melt pool at the substrate area is in good agreement with the experimental result. Specifically, the calculated contact-width between melt pool and substrate is 108 μm, while the one obtained from experiment of literature [3] is 120 μm. When it comes to the shape of the simulated melt pool's cross section, the result of the present study also demonstrates good agreement with the simulation results present in literatures [3, 4]. Consequently, the validity of the built up simulation model in the present study is confirmed.

3.5 Conclusion

This study has proposed a heat transfer simulation model considering the new volumetric heat source for SLM process. For taking into account the absorption of laser radiation by powder particles and substrate on the formation of the heat source in SLM, a developed sequential addition method to simulate the packed bed and Monte Carlo ray tracing simulation built up in Zemax are employed to construct the absorption profile along the depth of powder layer. Then, the equation for modeling the volumetric heat source is proposed. The validity of the proposed heat transfer model is confirmed by comparing the simulated result with experimental result from existing literature in terms of contact-width between melt pool and substrate. The proposed model can be further employed in the future to predict the stability of single scan track in SLM process.

Reference

1. Gibson, I., Rosen, D.W., Stucker, B.: Additive Manufacturing Technologies. Springer, New York (2010)
2. Foroozmehr, A., Badrossamay, M., Foroozmehr, E.: Finite element simulation of selective laser melting process considering optical penetration depth of laser in powder bed. Mater. Des. **89**, 255–263 (2016)
3. Gusarov, A., Yadroitsev, I., Bertrand, P., Smurov, I.: Model of radiation and heat transfer in laser-powder interaction zone at selective laser melting. J. Heat Transf. **131**, 072101 (2009)
4. Hodge, N., Ferencz, R., Solberg, J.: Implementation of a thermomechanical model for the simulation of selective laser melting. Comput. Mech. **54**, 33–51 (2014)

Chapter 4
Heat Conduction and Geometry Topology Optimization of Support Structure in Laser-Based Additive Manufacturing

Ehsan Malekipour, Andres Tovar, and Hazim El-Mounayri

Abstract Laser-based metal additive manufacturing technologies such as Selective Laser Sintering (SLS) and Selective Laser Melting (SLM) allow the fabrication of complex parts by selectively sintering or melting metallic powders layer by layer. Although elaborate features can be produced by these technologies, heat accumulation in overhangs leads to heat stress and warping, affecting the dimensional and geometrical accuracy of the part. This work introduces an approach to mitigate heat stress by minimizing the temperature gradient between the heat-accumulated zone in overhangs and the layers beneath. This is achieved by generating complex support structures that maintain the mechanical stability of the overhang and increase the heat conduction between these areas. The architecture of the complex support structures is obtained by maximizing heat conduction as an objective function to optimize the topology of support structure. This work examines the effect of various geometries on the objective function in order to select a suitable one to consume less material with almost same conduction. Ongoing work is the development of an experimental testbed for verification.

Keywords Additive manufacturing • Laser-based additive manufacturing • Support structure • Topology optimization • Heat stress

4.1 Introduction

Laser-based additive manufacturing (AM) is a manufacturing process in which a high power laser sinters/melts metal powders together [1–3]. Residual stress is one of the most frequent defect affecting significantly the precision of the process. The predominated reason for residual stress is heat stress because of temperature gradient existed in different zones [4–6]. This residual stress eventually leads to distortion and it may fail the whole process, which is significantly costly and consumes excess time. Nowadays manufacturers predominantly designs support structures to control distortion inside the part by increasing the resistance of the structure [7, 8]. This approach, as a trial and error method, significantly depends upon the designers' experiments [8]. Thus, first, there is no way to verify the efficiency of the designed support structure before fabrication of the part and second, there is no standard criteria/approach to follow for different geometries. As Fig. 4.1 shows, fabrication of support structures significantly effects on temperature uniformity, especially in overhangs [9] by increasing of heat conductivity and as a result, decreasing of temperature gradient. This specification of support structures can be employed efficiently to decrease heat stress in the fabricated part and eventually, reduces or eliminates distortion. More uniform temperature distribution also leads to more uniform microstructure. By generating more support structures (the dots in Fig. 4.1a show the positions of support structures), more uniform temperature distribution we can achieved (Fig. 4.1b) which means less temperature gradient. In addition, the temperature average decreases because of better conductivity [9].

In this regard, heat conduction in topology optimization is employed to design architecture of complex support structures. The top3d 169 lines code [10] is employed and modified here. The methodology is explained in the following section.

E. Malekipour (✉) • A. Tovar • H. El-Mounayri
Center for Additive Manufacturing Research at IUPUI (CAMRI), Purdue School of Engineering and Technology, Indianapolis, IN, USA
e-mail: emalekip@purdue.edu; tovara@iupui.edu; helmouna@iupui.edu

© The Society for Experimental Mechanics, Inc. 2018

J. Wang et al. (eds.), *Mechanics of Additive and Advanced Manufacturing, Volume 9*,
Conference Proceedings of the Society for Experimental Mechanics Series, DOI 10.1007/978-3-319-62834-9_4

4.2 Methodology

As it is explained in the introduction, there is a correlation between the number of fabricated support structures connecting overhang and its previous fabricated layer, and temperature distribution in a way that fabrication of enough number of support structures between an overhang and its previous layer, makes a temperature distribution uniform in the overhang layer [9]. If we assume we have enough number of support structures then, we have an overhang with a uniform temperature distribution. In this case, we can simplify the manufacturing model shown in Fig. 4.1a with a simple support structure in Fig. 4.2b. In Fig. 4.1a an overhang exerted by the laser beam (the heat affected zone (HAZ)) in its corner for fabricating of the next layer was joined to its previous layer of fabricated part (the heat sink) by a simple rectangular column as the support structure (Fig. 4.2a). Due to temperature uniformity on the overhang, we assume the same temperature on top face of support structure; simultaneously, the its bottom face will be the heat sink because it connected to the bottom layer which also reached to a uniform temperature by passing time.

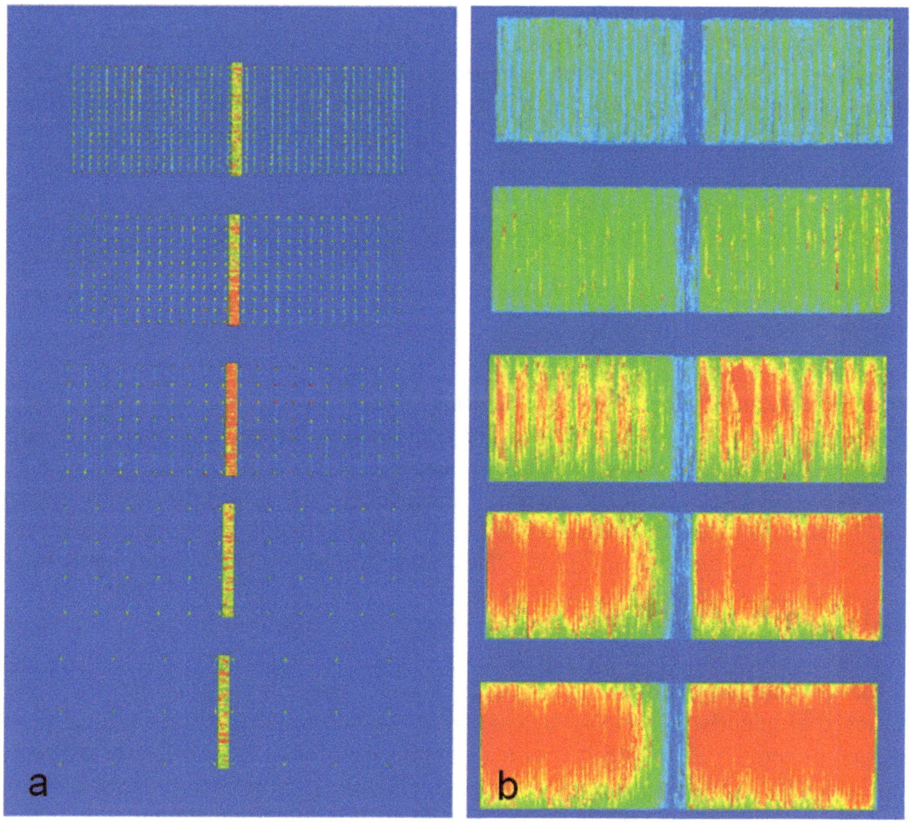

Fig. 4.1 Effect of support structure on temperature distribution and temperature gradient [9]

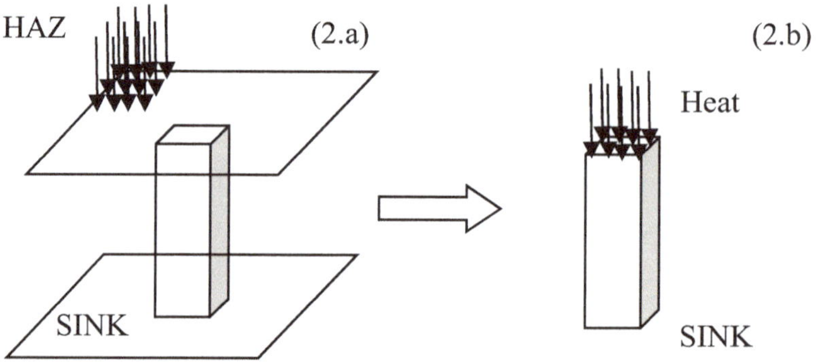

Fig. 4.2 Model of the problem (**a**) Real model with overhang, support structure and the layer beneath; (**b**) Simplified model

In this paper, we employs topology optimization to find the distribution of material in a prescribed area or volume referred to as the design domain by coding a binary programming problem introduced in reference [10]. In this regard, we employ topology optimization to maximize heat conduction in the designed support structures. Thus, manufacturers can fabricate the support structures with the same heat conductivity as the solid ones but with using fewer materials.

The equilibrium condition for heat transfer in finite element formulation is described by $\boldsymbol{K}(\boldsymbol{k}_i^0)\,\boldsymbol{U}(\boldsymbol{k}_i^0) = \boldsymbol{F}$ where $\boldsymbol{U}(\boldsymbol{k}_i^0)$ shows the finite element global nodal temperature vector, **F** shows the global thermal load vector, and $\boldsymbol{K}(\boldsymbol{k}_i^0)$ shows the global thermal conductivity matrix. The optimization problem for heat conduction is

$$\begin{aligned} &\text{Find} \quad \mathbf{K}_{\mathrm{i}}^{0} = [x_1, x_2, \ldots, x_e, \ldots, x_n]^T \\ &\text{Minimize} \quad \mathrm{c}(\mathbf{K}_{\mathrm{i}}^{0}) = \mathbf{F}^{\mathrm{T}}\mathbf{U}(\mathbf{k}_{\mathrm{i}}^{0}) \\ &\text{Subject to} \quad \nu(\tilde{\boldsymbol{x}}) = \tilde{\boldsymbol{x}}^T\mathbf{v} - \overline{\nu} \le 0 \\ &\qquad \mathbf{x} \in X, X = \{\mathbf{x} \in \mathrm{R}^{\mathrm{n}} : 0 \le \mathbf{x} \le 1\} \end{aligned} \tag{4.1}$$

Where $\mathbf{U}(\tilde{\mathbf{x}}) = \mathbf{K}(\tilde{\mathbf{x}})^{-1}\mathbf{F}$, and $\mathbf{K}(\tilde{\mathbf{x}})$ is obtained by the assembly of elements in thermal conductivity problem which is expressed as:

$$\boldsymbol{k}_i(\tilde{x_i}) = \left[\,k_{\min} + \tilde{x_i}^P(k_0 - k_{min})\right]k_i^0 \tag{4.2}$$

Where $k_{\min}$ and k_0 represent the limits of the material's thermal conductivity coefficient and k_i^0 shows the element conductivity matrix [10].

The primary top3d code solves minimum compliance for structural problems. By implementation of some modifications that are explained in following sentences, this code can also be used for heat conduction problem. The first modification is in heat conduction problem the number of DOF per node is one rather than three in minimum compliance in structure problems. The location of supports and applied forces also can be modified by considering the relationships between node number, node coordinates, node ID and node DOFs (Table 4.1).

Another modification is about creating various geometries by defining different active and passive elements. In this case, designers are able to calculate and compare the value of objective functions for different geometries and choose the support structure that consume fewer material but maximized objective function (heat conductivity) before the fabrication process. Objective function can be calculated by employing Optimality Criteria (OC) method. OC is a classical approach for solving a discretized structural optimization problem numerically. It is shown that this method is very efficient for solving the topology optimization problems [11].

The power law penalizes intermediate density values implicitly and make the resulting structure more black-and-white. This penalization procedure is usually referred to as the Solid Isotropic Material with Penalization (SIMP) method [12]. It should be mentioned that we consider the penalization constant and equal to 1.5 for all of the generated model in the future to eliminate the effect of varied penalization on the final results. A basic filter density function is defined as:

Table 4.1 Illustration of relationships between node number, node coordinates, node ID and node DOFs [10]

			Node degree of freedoms		
Node Number	Node coordinates	Node ID	x	y	z
N_1	(x_1, y_1, z_1)	NID_1[a]	$3 * \mathrm{NID}_1 - 2$	$3 * \mathrm{NID}_1 - 1$	$3 * \mathrm{NID}_1$
N_2	$(x_1 + 1, y_1, z_1)$	$\mathrm{NID}_2 = \mathrm{NID}_1 + (\mathrm{nely} + 1)$	$3 * \mathrm{NID}_2 - 2$	$3 * \mathrm{NID}_2 - 1$	$3 * \mathrm{NID}_2$
N_3	$(x_1 + 1, y_1 + 1, z_1)$	$\mathrm{NID}_3 = \mathrm{NID}_1 + \mathrm{nely}$	$3 * \mathrm{NID}_3 - 2$	$3 * \mathrm{NID}_3 - 1$	$3 * \mathrm{NID}_3$
N_4	$(x_1, y_1 + 1, z_1)$	$\mathrm{NID}_4 = \mathrm{NID}_1 - 1$	$3 * \mathrm{NID}_4 - 2$	$3 * \mathrm{NID}_4 - 1$	$3 * \mathrm{NID}_4$
N_5	$(x_1, y_1, z_1 + 1)$	$\mathrm{NID}_5 = \mathrm{NID}_1 + \mathrm{NID}_z$[b]	$3 * \mathrm{NID}_5 - 2$	$3 * \mathrm{NID}_5 - 1$	$3 * \mathrm{NID}_5$
N_6	$(x_1 + 1, y_1, z_1 + 1)$	$\mathrm{NID}_6 = \mathrm{NID}_2 + \mathrm{NID}_z$	$3 * \mathrm{NID}_6 - 2$	$3 * \mathrm{NID}_6 - 1$	$3 * \mathrm{NID}_6$
N_7	$(x_1 + 1, y_1 + 1, z_1 + 1)$	$\mathrm{NID}_7 = \mathrm{NID}_3 + \mathrm{NID}_z$	$3 * \mathrm{NID}_7 - 2$	$3 * \mathrm{NID}_7 - 1$	$3 * \mathrm{NID}_7$
N_8	$(x_1, y_1 + 1, z_1 + 1)$	$\mathrm{NID}_8 = \mathrm{NID}_4 + \mathrm{NID}_z$	$3 * \mathrm{NID}_8 - 2$	$3 * \mathrm{NID}_8 - 1$	$3 * \mathrm{NID}_8$

[a] $\mathrm{NID}_1 = z_1*(\mathrm{nelx} + 1)*(\mathrm{nely} + 1) + x_1*(\mathrm{nely} + 1) + (\mathrm{nely} + 1 - y_1)$

[b] $\mathrm{NID}_z = (\mathrm{nelx} + 1)*(\mathrm{nely} + 1)$

$$x_i^{\sim} = \frac{\sum_{j \in N_i} H_{ij} \nu_j x_j}{\sum_{j \in N_i} H_{ij} \nu_j} \tag{4.3}$$

Where N_i is the neighborhood of an element x_i with volume ν_i, and H_{ij} is a weight factor. The filtered density $x_i^{\sim}$ indicates a modified (physical) density field. The modified conductivity used for heat conduction problem is:

$$\boldsymbol{k}_i(\widetilde{x_i}) = \left[k_{\min} + \widetilde{x_i}_i^{\,p}(k_0 - k_{min}) \right] k_i^0 \; x_i^{\sim} \in [0, 1] \tag{4.4}$$

It should be noted that the traditional direct solver does not work expeditiously for the finite element large-scale problems. Nevertheless, iterative solver can solve large-scale problems efficiently [13, 14]. In this regard, line 72 of top3d code is replaced by a built-in MATLAB function pcg, called preconditioned conjugate gradients method.

4.3 Result and Discussion

As it was already explained the main objective is to optimize the topology of a simple rectangular column as a support structure between an overhang and its previous layer. As Fig. 4.3a shows, the primary model in top3d code for heat conduction includes a rectangular cube, which has a sink on the middle of top surface, and all other nodes are given a thermal load.

Fig. 4.3b shows the resulting shape for this case. In the next step, we change the geometry with using different pattern of active and passive elements (see Appendix, section: DEFINING PASSIVE ELEMENTS Status, line 70 to 131) to find the geometry generating minimum objective function and as a result, maximum heat conductivity. Finally, the boundary conditions (heat flux and sink) are applied on the obtained geometry to model the desired laser-based AM process. It is noteworthy that for eliminating the effect of other parameters on the final result, we keep the volume fraction equals to 0.4, penalization equals to 3, minimum radius filter (rmin) equals to 1.5, number of iteration equals to 60 and dimensions equals to 80 × 80 × 20 fixed for all tests.

Table 4.2 shows that the value of objective function is minimum for the geometry without any passive elements (any holes). This result is predictable because solid support structure conducts heat better than any other types of geometry. But the best option is the fourth one since the criteria for choosing the most optimized support structure is the one with minimum objective function (maximum conductivity) while consume less material. As Table 4.2 shows the percentage difference between the value of objective functions of number one and four is just around 11% while it uses less material. Thus, we choose the geometry type number four as the best one for support structure; The objective function variation for different types of geometry is revealed in the following diagram (Fig. 4.4).

As the final step, the desired boundary conditions (exerted heat and sink zone) are applied on the chosen geometry (see Appendix, section: Different load conditions, line 24 to 29) and the resulting shape of support structure are derived. This shape is the optimized support structure that should be used by vendors in the specified manufacturing conditions (Table 4.3).

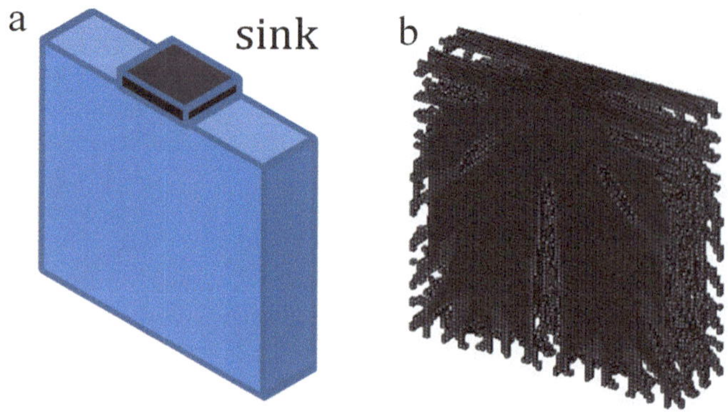

Fig. 4.3 (**a**) The primary model for support structure; (**b**) The resulting topology optimized shape

Table 4.2 Generating various geometries by using different patterns of active and passive elements

No.	Geometrical type	Value of objective function	Resulting shape
1	Primary model	Obj.: 97580.9442	
2	Primary model set limits on two sides	Obj.:189254.3599	
3	Primary model with a big hole at the middle	Obj.:332743.6824	
4	Primary model with a small hole at the middle	Obj.:109929.4235	
5	Primary model with a small hole at the middle top	Obj.:134462.7200	
6	Primary model with two small holes at the middle top and middle bottom	Obj.:148970.7845	

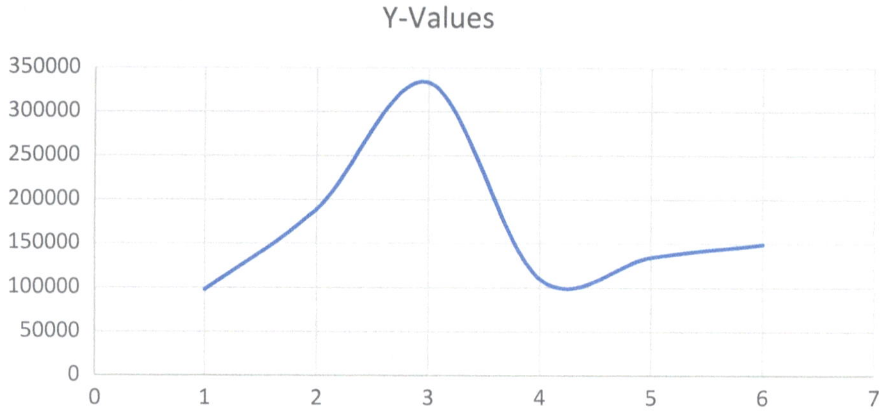

Fig. 4.4 The value of objective function for different types of geometry revealed in Table 4.2

Table 4.3 Optimized support structure for different boundary conditions

No.	Boundary conditions	Value of objective function	Optimized support structure
1	Sink is placed at the bottom face	Obj.:109614.2276	
2	Heat flux is exerted on half of the top face and sink is placed at the bottom face	Obj.:274029098.3609	
3	Heat flux is exerted on the top face and sink is placed at the bottom face	Obj.:431901.2664	
4	Heat flux is exerted on one third of the top face and sink is placed at the bottom face	Obj.: 46243.4644	

4.4 Summery and Conclusions

This paper shows how manufacturers can optimize support structures topologically between overhangs and the layer beneath, in order to maximize heat conduction subjected to structural stiffness constraint in laser-based AM.

In this regard, top3d 169 lines MATLAB code is modified and employed. Firstly, various geometries are generated by defining different patterns for active and passive elements and then, the best geometry is identified based on maximum heat conduction while using less material. The results show that the value of objective function for the geometry with a small hole in the middle of structure is minimum showing maximum heat conduction. Finally, different boundary conditions with respect to different fabrication circumstances are applied and the final topological optimized shape for support structures are derived. It should be noticed that in this paper we just examined four different types of geometry (Table 4.2) and also, utilize mono-physical objective function. As the next step, we are going to examine more types of geometries and also, utilize multi-physics objective function in which heat conduction is maximized subjected to structural stiffness constraint and finally, development of an experimental testbed for verification.

References

1. Gu, D., Meiners, W., Wissenbach, K., Poprawe, R.: Laser additive manufacturing of metallic components: materials, processes and mechanisms. Int. Mater. Rev. **57**, 133–164 (2012)
2. Bhavar, V., Kattire, P., Patil, V., Khot, S., Gujar, K., Singh, R.: A review on powder bed fusion technology of metal additive manufacturing. In: International Conference & Exhibition on Additive Manufacturing Technologies Bangalore, India (2015)
3. Frazier, W.E.: Metal additive manufacturing: a review. J. Mater. Eng. Perform. **23**, 1917–1928 (2014)
4. Zaeh, M.F., Branner, G.: Investigations on residual stresses and deformations in selective laser melting. Prod. Eng. **4**, 35–45 (2010)
5. Roberts, I.A.: Investigation of residual stresses in the laser melting of metal powders in additive layer manufacturing, University of Wolverhampton, Netherlands (2012)
6. Cheng, B., Chou, K.: Thermal stresses associated with part overhang geometry in electron beam additive manufacturing: process parameter effects. Proc. Annu. Int. Solid Freeform Fabr. Symp. 1076 (2014)
7. Krol, T., Zäh, M., Seidel, C.: Optimization of supports in metal-based additive manufacturing by means of finite element models. Bourell DL (2012)
8. Calignano, F.: Design optimization of supports for overhanging structures in aluminum and titanium alloys by selective laser melting. Mater. Des. **64**, 203–213 (2014)
9. Craeghs, T., Clijsters, S., Kruth, J.-P., Bechmann, F., Ebert, M.-C.: Detection of process failures in layerwise laser melting with optical process monitoring. Phys. Procedia. **39**, 753–759 (2012)
10. Liu, K., Tovar, A.: An efficient 3D topology optimization code written in Matlab. Struct. Multidiscip. Optim. **50**, 1175–1196 (2014)
11. Christensen, P.W., Klarbring, A.: An Introduction to Structural Optimization, vol. 153. Springer Science & Business Media, Berlin (2008)
12. Zhou, M., Rozvany, G.: The COC algorithm, part II: topological, geometrical and generalized shape optimization. Comput. Methods Appl. Mech. Eng. **89**, 309–336 (1991)
13. Hestenes, M.R., Stiefel, E.: Methods of Conjugate Gradients for Solving Linear Systems, vol. 49. NBS, Washington, DC (1952)
14. Augarde, C., Ramage, A., Staudacher, J.: An element-based displacement preconditioner for linear elasticity problems. Comput. Struct. **84**, 2306–2315 (2006)

Appendix

```
% AN 169 LINE 3D TOPOLOGY OPITMIZATION CODE BY LIU AND TOVAR (JUL 2013)
function top3d(volfrac,penal,rmin)   % Guide: enter this in Command Window: top3d(0.4,3.0,1.5)
% USER-DEFINED LOOP PARAMETERS
maxloop = 60;    % Maximum number of iterations
tolx = 0.01;     % Terminarion criterion
displayflag = 0; % Display structure flag
% GEOMETRY DESIGN
nelx = 80;
nely = 70;
nelz = 20;
% USER-DEFINED MATERIAL PROPERTIES
k0 = 1;          % Good thermal Conductivity
kmin = 1e-3;     % Poor thermal Conductivity
% USER-DEFINED SUPPORT FIXED DOFs
il = nelx/2-nelx/20:nelx/2+nelx/20; jl = 0; kl = 0:nelz;
fixedxy = il*(nely+1)+(nely+1-jl);
fixednid = repmat(fixedxy',size(kl))+ ...
  repmat(kl*(nelx+1)*(nely+1),size(fixedxy,2),1);
fixeddof = reshape(fixednid,[],1);
% PREPARE FINITE ELEMENT ANALYSIS
nele = nelx*nely*nelz;
ndof = (nelx+1)*(nely+1)*(nelz+1);
% F = sparse(1:ndof,1,-0.01,ndof,1);
% Different load conditions
F = zeros(ndof,1);
% F(1:ndof/2,1) = 1;   % half of nodes
% F(nely+1:nely+1:ndof,1) = 1;   % bottom face
% F(1:nely+1:ndof,1) = 1;   % top face
F(1:nely+1:ndof/3,1) = 1;   % one third of top face
U = zeros(ndof,1);
freedofs = setdiff(1:ndof,fixeddof);
KE = lk_H8(k0);
nodegrd = reshape(1:(nely+1)*(nelx+1),nely+1,nelx+1);
nodeids = reshape(nodegrd(1:end-1,1:end-1),nely*nelx,1);
nodeidz = 0:(nely+1)*(nelx+1):(nelz-1)*(nely+1)*(nelx+1);
nodeids = repmat(nodeids,size(nodeidz))+repmat(nodeidz,size(nodeids));
edofVec = nodeids(:)+1;
edofMat = repmat(edofVec,1,8)+ ...
  repmat([0 nely + [1 0] -1 ...
  (nely+1)*(nelx+1)+[0 nely + [1 0] -1]],nele,1);
iK = reshape(kron(edofMat,ones(8,1))',8*8*nele,1);
jK = reshape(kron(edofMat,ones(1,8))',8*8*nele,1);
% PREPARE FILTER
iH = ones(nele*(2*(ceil(rmin)-1)+1)^2,1);
jH = ones(size(iH));
sH = zeros(size(iH));
k = 0;
for k1 = 1:nelz
  for i1 = 1:nelx
    for j1 = 1:nely
      e1 = (k1-1)*nelx*nely + (i1-1)*nely+j1;
      for k2 = max(k1-(ceil(rmin)-1),1):min(k1+(ceil(rmin)-1),nelz)
        for i2 = max(i1-(ceil(rmin)-1),1):min(i1+(ceil(rmin)-1),nelx)
          for j2 = max(j1-(ceil(rmin)-1),1):min(j1+(ceil(rmin)-1),nely)
            e2 = (k2-1)*nelx*nely + (i2-1)*nely+j2;
            k = k+1;

            iH(k) = e1;
            jH(k) = e2;
            sH(k) = max(0,rmin-sqrt((i1-i2)^2+(j1-j2)^2+(k1-k2)^2));
          end
        end
      end
    end
  end
end
```

```
H = sparse(iH,jH,sH);
Hs = sum(H,2);
% INITIALIZE ITERATION
x = repmat(volfrac,[nely,nelx,nelz]);
% DEFINING PASSIVE ELEMENTS Status
% A. an small hole inside middle
for ely = 1:nely
  for elx = 1:nelx
    if sqrt((ely-nely/2.)^2+(elx-nelx/2.)^2) < nely/6.
      passive(ely,elx) = 1;
    else
      passive(ely,elx) = 0;
    end
  end
end

passive = repmat(passive,[1,1,nelz]);
x (find(passive)) = 0;
%
% B. two small holes
% for ely = 1:nely
%   for elx = 1:nelx
%     if sqrt((ely-nely/4.)^2+(elx-nelx/2.)^2) < nely/6.
%       passive1(ely,elx) = 1;
%     else
%       passive1(ely,elx) = 0;
%
%     end
%   end
% end
%
% for ely = 1:nely
%   for elx = 1:nelx
%       if sqrt((ely-3*(nely/4.))^2+(elx-nelx/2.)^2) < nely/6.
%       passive2(ely,elx) = 1;
%     else
%       passive2(ely,elx) = 0;
%     end
%   end
% end
%
% passive1 = repmat(passive1,[1,1,nelz]);
% passive2 = repmat(passive2,[1,1,nelz]);
%
% x (find(passive1)) = 0;
% x (find(passive2)) = 0;

% C. two limited face
% for ely = 1:nely
%   for elx = 1:10
%     passive(ely,elx) = 1;
%   end
%
%
%   for elx = 10:70
%   passive(ely,elx) = 0;
%   end
%
%   for elx = 70:80
%   passive(ely,elx) = 1;
%   end
% end
%
% passive = repmat(passive,[1,1,nelz]);
% x (find(passive)) = 0;
%
```

```
xPhys = x;
loop = 0;
change = 1;
% START ITERATION
while change > tolx && loop < maxloop
  loop = loop+1;
  % FE-ANALYSIS
  sK = reshape(KE(:)*(kmin+(1-kmin)*xPhys(:)'.^penal),8*8*nele,1);
  K = sparse(iK,jK,sK); K = (K+K')/2;
%   U(freedofs,:) = K(freedofs,freedofs)\F(freedofs,:);
  % For large mesh size
  tolit = 1e-8;
  maxit = 8000;
  M = diag(diag(K(freedofs, freedofs)));
  U(freedofs,:) = pcg(K(freedofs, freedofs) ,F(freedofs,:),tolit,maxit,M);
  % OBJECTIVE FUNCTION AND SENSITIVITY ANALYSIS
  ce = reshape(sum((U(edofMat)*KE).*U(edofMat),2),[nely,nelx,nelz]);
  % active the following line for one hole
  xnew(find(passive)) = 0;
  % active these two lines for two small holes command
%   xnew(find(passive1)) = 0;
%   xnew(find(passive2)) = 0;
  c = sum(sum(sum((kmin+(1-kmin)*xPhys.^penal).*ce)));
  dc = -penal*(1-kmin)*xPhys.^(penal-1).*ce;
  dv = ones(nely,nelx,nelz);
  % FILTERING AND MODIFICATION OF SENSITIVITIES
  dc(:) = H*(dc(:)./Hs);
  dv(:) = H*(dv(:)./Hs);
  % OPTIMALITY CRITERIA UPDATE
  l1 = 0; l2 = 1e9; move = 0.2;
  while (l2-l1)/(l1+l2) > 1e-3
    lmid = 0.5*(l2+l1);
    xnew = max(0,max(x-move,min(1,min(x+move,x.*sqrt(-dc./dv/lmid)))));
    xPhys(:) = (H*xnew(:))./Hs;
    if sum(xPhys(:)) > volfrac*nele, l1 = lmid; else l2 = lmid; end
  end
  change = max(abs(xnew(:)-x(:)));
  x = xnew;
  % PRINT RESULTS
  fprintf(' It.:%5i Obj.:%11.4f Vol.:%7.3f ch.:%7.3f\n',loop,c,mean(xPhys(:)),change);
  % PLOT DENSITIES
  if displayflag, clf; display_3D(xPhys); end %#ok<UNRCH>
end
clf; display_3D(xPhys);
end


% === GENERATE ELEMENT STIFFNESS MATRIX ===
function [KE] =lk_H8(k);
A1 = 4*eye(2); A2=-eye(2);
A3 = fliplr(A2); A4 = -ones(2);
KE1 = [A1 A2; A2 A1];
KE2 = [A3 A4; A4 A3];
KE= 1/12 * k * [KE1 KE2; KE2 KE1];
end
% === DISPLAY 3D TOPOLOGY (ISO-VIEW) ===
function display_3D(rho)
[nely,nelx,nelz] = size(rho);
hx = 1; hy = 1; hz = 1;            % User-defined unit element size
face = [1 2 3 4; 2 6 7 3; 4 3 7 8; 1 5 8 4; 1 2 6 5; 5 6 7 8];
set(gcf,'Name','ISO display','NumberTitle','off');
for k = 1:nelz
  z = (k-1)*hz;
  for i = 1:nelx
    x = (i-1)*hx;
    for j = 1:nely
```

```
                y = nely*hy - (j-1)*hy;
                if (rho(j,i,k) > 0.5)  % User-defined display density threshold
                    vert = [x y z; x y-hx z; x+hx y-hx z; x+hx y z; x y z+hx;x y-hx z+hx; x+hx y-hx z+hx;x+hx y z+hx];
                    vert(:,[2 3]) = vert(:,[3 2]); vert(:,2,:) = -vert(:,2,:);
                    patch('Faces',face,'Vertices',vert,'FaceColor',[0.2+0.8*(1-rho(j,i,k)),0.2+0.8*(1-rho(j,i,k)),0.2+0.8*(1-rho(j,i,k))]);
                    hold on;
                end
            end
        end
    end
    axis equal; axis tight; axis off; box on; view([30,30]); pause(1e-6);
    end
    % =================================================================================
    % === This code was written by K Liu and A Tovar, Dept. of Mechanical      ===
    % === Engineering, Indiana University-Purdue University Indianapolis,      ===
    % === Indiana, United States of America                                    ===
    % === ---------------------------------------------------------------      ===
    % === Please send your suggestions and comments to: kailiu@iupui.edu       ===
    % === ---------------------------------------------------------------      ===
    % === downloaded from the website: http://www.top3dapp.com/                ===
    % === ---------------------------------------------------------------      ===
    % === Disclaimer:                                                          ===
    % === The authors reserves all rights for the program.                     ===
    % === The code may be distributed and used for educational purposes.       ===
```

Chapter 5
Strain Energy Dissipation Mechanisms in Carbon Nanotube Composites Fabricated by Additive Manufacturing

Frank Gardea, Daniel Cole, Bryan Glaz, and Jaret Riddick

Abstract Carbon nanotube (CNT) reinforced acrylonitrile-butadiene-styrene (ABS) composites fabricated using a fused deposition modeling approach were characterized for mechanical strain energy storage and dissipation capabilities. Dynamic mechanical analysis (DMA) was performed to quantify loss factor as a function of applied dynamic strain. In addition, DMA was performed at varying temperatures to give insight into the molecular interactions present in these composites. Insight into the microstructure was provided by atomic force microscopy (AFM). The results are compared to neat ABS and ABS/CNT systems processed through an injection molding technique. Results show large energy dissipation, accompanied by permanent damage, in both injection molded and additively manufactured neat ABS samples. In contrast, the additively manufactured ABS/CNT nanocomposites exhibited strain energy dissipation but reduced the effect of cavitation and crazing by possible reinforcement. This result suggests CNT fillers have the potential to alter the dissipation mechanisms present in additively manufactured structures to control structural damping. This research provides insight into the design and additive manufacturing of materials where energy dissipation is essential to maintain structural stability and functionality under dynamic loading.

Keywords Additive manufacturing • Damping • Nanocomposites • Carbon nanotubes • Energy dissipation

5.1 Introduction

Polymer based composites consisting of nanoscale fillers show promise for enabling multifunctional nanocomposites with unique properties. The new possibilities range from enhanced mechanical, electrical, and thermal properties to functionalities such as self-healing, sensing, and power storage. However, even as advanced manufacturing methods are developed, the promise of unique materials with advanced capabilities has not been fully met. An advanced manufacturing method that has gained a significant amount of attention is additive manufacturing (AM). Additive manufacturing consists of building a product layer-by-layer, one cross-section at a time, rather than using the conventional method of subtracting material from a larger piece of material [1]. While additive manufacturing enables many unique capabilities, such as geometrical freedom in engineering design, on-demand fabrication, and functional parts without the need for assembly, it also comes with limitations, which include poor material properties of printed parts as compared to conventional manufacturing processes and a narrow selection of available materials, with most being proprietary. In addition, the lack of characterization and knowledge of these materials limits the applications to non-load bearing products and prototypes.

The merging of AM with nanotechnology presents a new approach to addressing and overcoming the limitations present by the use of fundamental materials and design limitations. Nanofillers have the potential to alleviate some of these limitations, by introducing reinforcement and multifunctionality to the manufactured materials. The printing of nanocomposites allows for the development of new printed multifunctional materials, and provides the potential for the control of material properties at the nanoscale. Studies have been done on AM of polymer filaments filled with micron sized fibers (i.e. short carbon fiber) [2–4], however there are limited publications on the use of nanomaterials in AM materials/polymers. However, some recent studies on AM have shown progress on realizing improved global mechanical properties through the addition of nanofillers. Shofner et al. [5] fabricated nanofiber-reinforced polymers using fused deposition modeling and found that the nanofibers increased the tensile strength and elastic modulus by 39% and 60%, respectively. However, strain energy dissipation mechanisms were not investigated. Wahab et al. [6] used nanoparticles of zirconia and

F. Gardea (✉)
Vehicle Technology Directorate, U.S. Army Research Laboratory, Aberdeen Proving Ground, MD 21005, USA
e-mail: frank.gardea4.civ@mail.mil

D. Cole • B. Glaz • J. Riddick
Vehicle Technology Directorate, U.S. Army Research Laboratory, Aberdeen Proving Ground, Aberdeen, MD 21005, USA

© The Society for Experimental Mechanics, Inc. 2018

J. Wang et al. (eds.), *Mechanics of Additive and Advanced Manufacturing, Volume 9*,
Conference Proceedings of the Society for Experimental Mechanics Series, DOI 10.1007/978-3-319-62834-9_5

clay to fabricate zirconia/polyamide and clay/polyamide composites using a selective laser sintering technique. However, the mechanical properties decreased as compared to the neat polymer which were attributed to a large number of voids within the material. A conductive CNT/poly(lactic acid) composite was successfully fabricated by Postiglione et al. [7] using liquid deposition modeling. Compton and Lewis [8] developed an epoxy-based ink with high aspect ratio silicon carbide whiskers and milled carbon fibers as reinforcing agents. By tailoring the composition and rheology, the printability of the ink and the orientation of the fibers was controlled. Increases in elastic modulus of up to one order of magnitude as compared to commercial thermoplastics and photocurable resins was observed.

Nanofillers have been used to enhance mechanical properties of polymer composites where efficient stress transfer, with limited or no sliding, at the interface leads to increases in strength and stiffness [9, 10]. However, relative sliding of the filler and matrix can lead to mechanical energy dissipation as a result of frictional sliding. This damping mechanism has been shown to be effective in dissipating mechanical energy by Suhr and Koratkar [11], in which a CNT/polycarbonate system showed 4–5 times increase in loss modulus relative to the neat polycarbonate matrix. Gardea et al. [12] showed that the nature of interfacial slip in aligned CNT/PS composites was due to mechanical interlocking. Ogasawara et al. [13] studied CNT/PEEK composites and showed hysteretic behavior in these composites, which was attributed to interfacial slip between the CNT and polymer matrix. Even though studies show improved damping properties and shed light into the mechanisms behind the increased dissipation, there is currently no understanding of the fundamental interfacial mechanics of AM based nanocomposites.

The current work attempts to address the issue of limited knowledge of interfacial mechanics of AM based nanocomposites by studying the effect of the AM process on the material response and microstructure of CNT loaded ABS matrix composites. In addition, the effect of the CNTs on the dynamic properties of these AM samples is investigated. A comparison is made to the dissipation mechanisms of nanocomposites from the traditional fabrication process of injection molding. This work also addresses the issue of limited material selection by developing a printable nanocomposite material which can lead to expansion of the application of AM products.

5.2 Experimental

5.2.1 *Processing*

Multi-walled carbon nanotubes with an outer diameter of 30–50 nm and length of 10–20 μm from Cheap Tubes Inc. were used with no surface functionalization. The matrix material used was Lustran® 633 acrylonitrile-butadiene-styrene (ABS) polymer. The fabrication of the composites consisted of first dissolving the ABS pellets in toluene and directly mixing the CNTs (1 wt%). The toluene was then evaporated to obtain a batch of ABS/CNT material. This material was then mixed using a twin-screw compounder for a total of 30 min at 100 rpm and at a temperature of 210 °C. The mixed material was either injection molded into a square mold and cut into samples for DMA testing, or extruded through a circular die. The extruded material was then processed through a second extruder to obtain a filament with the appropriate diameter for printing. The additive manufacturing was conducted using a Fused Deposition Modeling (FDM) approach with the use of a MakerBot Replicator 2X experimental printer. The printing was performed at a nozzle temperature of 230 °C and a base temperature of 110 °C. All samples where printed in a unidirectional direction with all layers having the builds in the same [90°] direction, as observed in Fig. 5.1. All mechanical testing coincided with loading with respect to this printing direction.

5.2.2 *Characterization*

Dynamic mechanical analysis was performed using a single cantilever beam set-up in a TA instruments Q800 DMA system at room temperature. A strain amplitude sweep of 0.01–3% was performed. The frequency was held constant at 1.0 Hz during all tests. The accumulated damage was analyzed by performing a second strain sweep from 0.01% to 3%.

Temperature dependent viscoelastic properties were analyzed at an applied tensile strain of 0.1%, a preload of 0.01 N, and a force tracking of 125%. A temperature sweep from −145 to 120 °C was applied at a rate of 2 °C/min. To study the strain recovery, and thus the possible polymer chain alignment in the samples, a temperature ramp was conducted at a constant stress of 4 kPa, as shown in Fig. 5.2, where the max temperature of 120 °C (above the glass transition, T_g, of the polymer) was held constant and the material allowed to contract or expand.

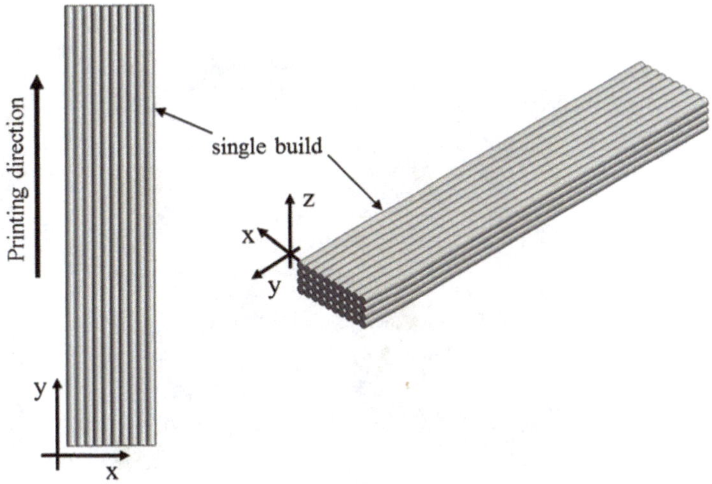

Fig. 5.1 Schematic showing the printing direction and [90°] orientation of each build composing the AM sample

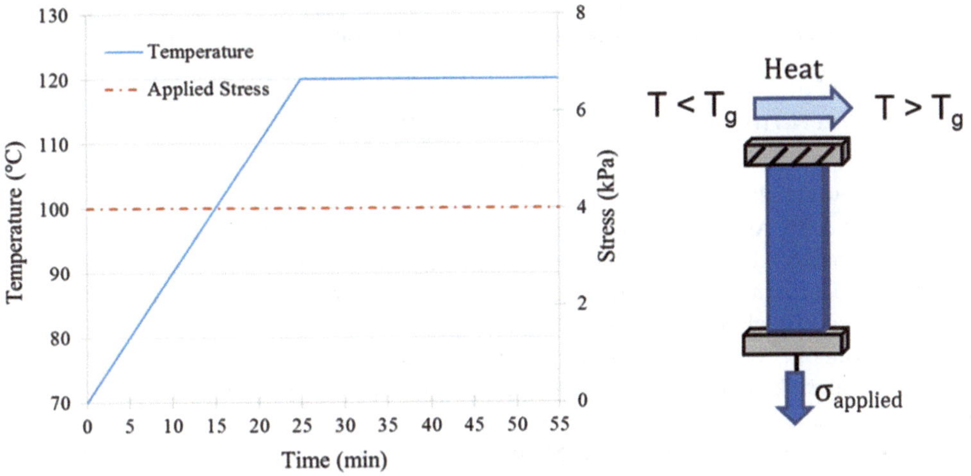

Fig. 5.2 Applied stress, temperature profile and schematic for the free recovery test in which the sample is allowed to expand or contract when heated above the T_g of the polymer

Atomic force microscopy (AFM) was performed on cryogenically microtomed surfaces in order to study the effects of injection molding and additive manufacturing on the resulting microstructure. An Asylum Research Cypher atomic force microscope was used in alternating contact mode. The AFM probes used in this study were Al-coated Si cantilevers with nominal natural frequency, stiffness, and tip radius of curvature of ~115 kHz, ~9 N/m, and ~10 nm, respectively.

5.3 Results and Discussion

Results of the interfacial mechanics present in both injection molded and AM nanocomposites are presented. First, the effect of the manufacturing process on the microstructure is shown by AFM analysis and strain recovery results. The impact of the manufacturing process on the dynamic mechanical properties, specifically the storage modulus and damping factor tan δ, of neat ABS is then presented. Finally, the effect on dynamic behavior by the addition of CNT reinforcements to the ABS matrix is discussed.

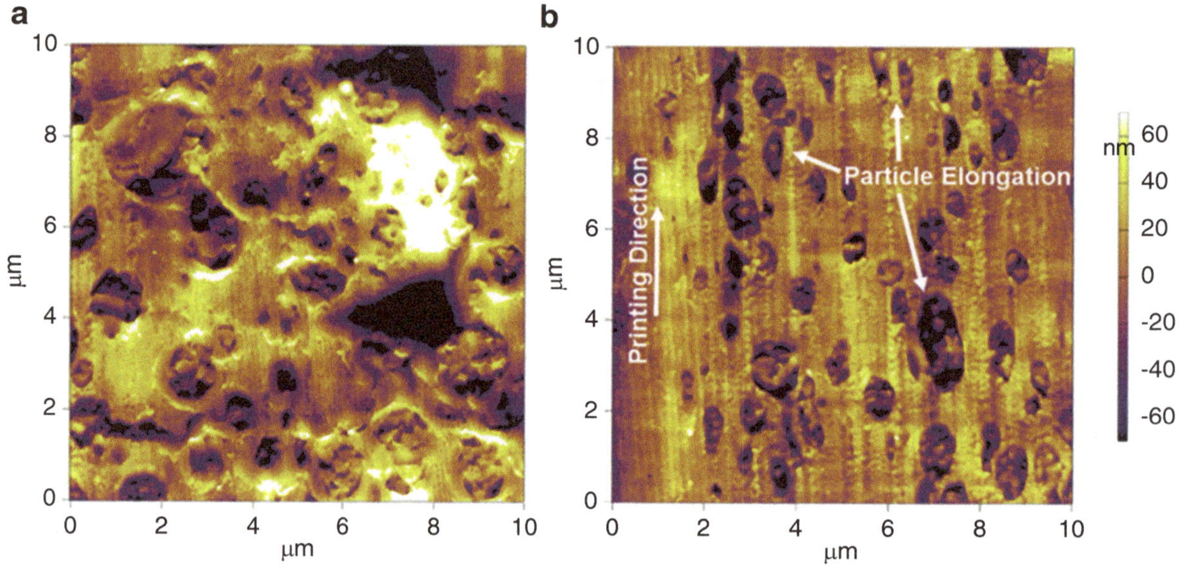

Fig. 5.3 AFM height scans of microtomed neat ABS polymer samples fabricated by (**a**) injection molding and (**b**) additive manufacturing

5.3.1 Effect of Manufacturing Process on Microstructure

To observe if any alignment was induced by the AM process, AFM was conducted on microtomed surfaces of the samples. As seen in Fig. 5.3, the neat ABS AM sample shows elongation of the butadiene particles in the direction of printing, as compared to the injection molded ABS which shows the butadiene particles maintaining a relatively spherical shape. This elongation suggests that the alignment present in AM samples is due to the shear induced by the travel of the printing nozzle. AM CNT composites show similar elongation of the butadiene particles. This is an important result for understanding and controlling failure mechanisms in AM materials since the rubbery butadiene particles are present in order to increase the fracture toughness of the material. A preferred butadiene phase orientation could potentially be manipulated in order to manipulate fracture in a particular direction.

In order to verify whether there was any programmed strain in the AM samples resulting from polymer chain alignment during the AM process, a free recovery (small applied stress) test was conducted above the T_g of the polymer. The free recovery experiments were conducted to determine the maximum recovered extension. As observed in Fig. 5.4, the AM process induced polymer chain alignment in both neat ABS and CNT/ABS composite samples. This alignment is evident in the present strain recovery (~25.7%) of the AM neat ABS, as opposed to the expansion (~34.3%) observed in the injection molded neat ABS samples. The CNT composites show less recovery (~21.3%) and expansion (~28.2%), suggesting the presence of the CNTs restricts further relaxation of the polymer chains, which supports the notion that the CNTs act as reinforcing agents within the surrounding matrix and butadiene particles.

5.3.2 Dynamic Mechanical Properties of Neat ABS

When comparing the initial storage modulus between injection molded and AM neat ABS samples, it is observed that the AM process results in an increase in the storage modulus of printed parts of ~6.1%, Fig. 5.5a. This result suggests that the AM process produces a reinforcing effect in the direction of printing, which is explained by induced alignment of the polymer chains. The uncoiling of the polymer chains by the printing process allows the material to sustain higher stresses in the direction of alignment. It is understood that the main dissipation mechanisms present in neat ABS are the cavitation of the butadiene rubber particles and crazing in the matrix initiated at the rubber/polymer interface [14–16]. These mechanisms lead to significant energy dissipation, accompanied by significant decrease in the storage modulus as observed in the dynamic results of the injection molded neat ABS samples in Fig. 5.5a and b. Upon reloading (cycle 2) significant accumulated damage is observed, suggesting significant cavitation and crazing is present in the samples. The effect of chain alignment is also evident in the damping characteristics of the material. As observed in Fig. 5.5b, at low strains in the

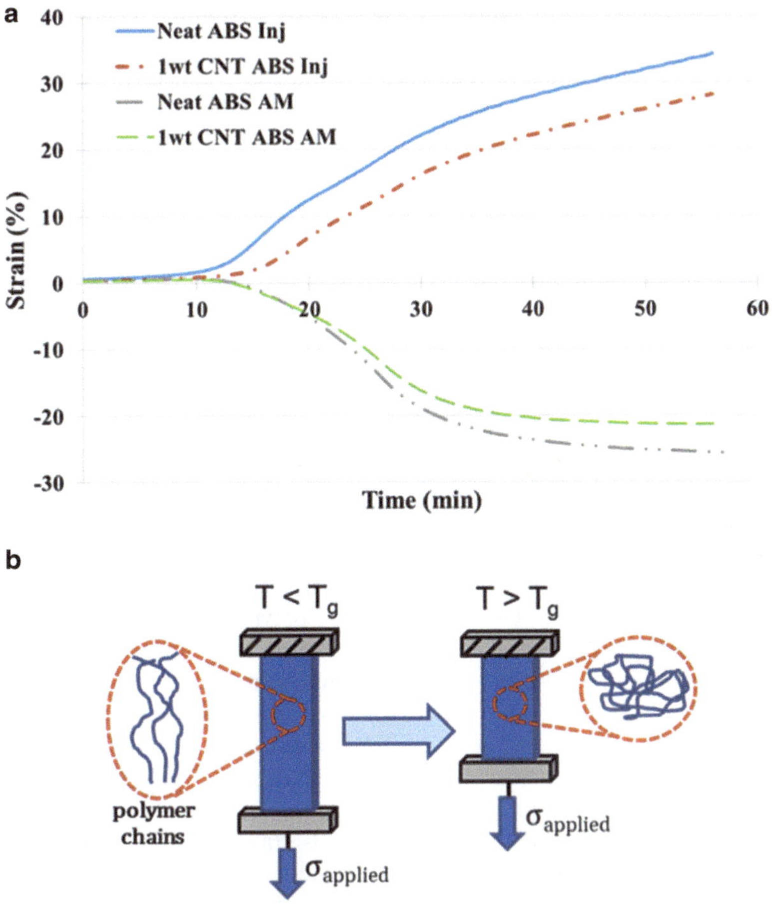

Fig. 5.4 (**a**) Free recovery experiments conducted on both injection molded and additively manufactured samples for both neat ABS and 1 wt% CNT/ABS composites (**b**) schematic showing the alignment of polymer chains below T_g and the strain recovery induced by the relaxation of the polymer chains above T_g

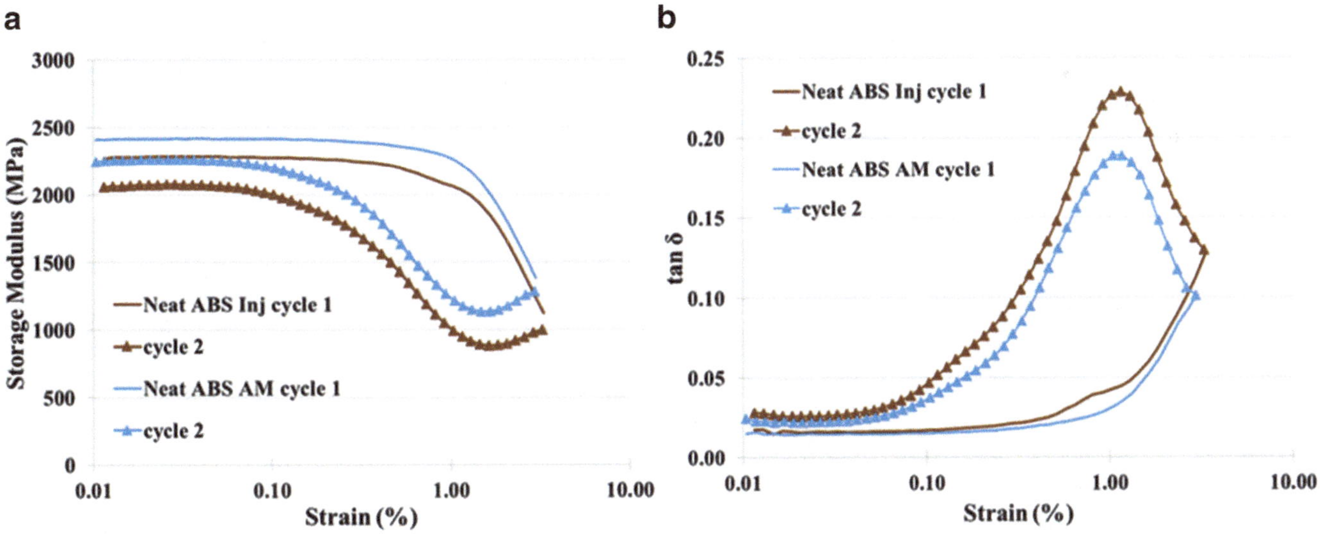

Fig. 5.5 (**a**) Storage modulus and (**b**) tan δ as a function of applied strain for neat ABS fabricated by injection molding (Inj) and additive manufacturing (AM)

first cycle, the loss moduli are relatively similar (due to polymer-polymer interaction). As the applied strain increases, dissipation in the both samples increases as the crazing and cavitation mechanisms are triggered. On the second cycle, the AM neat ABS samples show a decrease of ~17.4% and narrowing of the tan δ peak compared to the injection molded sample. Although less dissipation is present within the material, these results suggest that the primary dissipation mechanisms remain the same since there in no shift in the location of the peak. Some possible explanations for the reduced tan δ of the AM samples compared to the injection molded are that cavitation and crazing are suppressed by the alignment of the polymer chains, possibly due to a constraint of the butadiene particles and a decrease in the formation of crazes resulting from decreased entanglement density.

5.3.3 Dynamic Mechanical Properties of CNT/ABS Nanocomposites

The CNTs present in the ABS matrix act as reinforcing agents, as observed in the increase in storage modulus of ~9.7% between neat ABS and 1 wt% CNT/ABS nanocomposites fabricated by injection molding, shown in Fig. 5.6a. The change in modulus in the CNT composite shows the same trend as in the neat ABS samples, suggesting that both materials have the same dissipation mechanisms and no additional mechanisms are introduced by the presence of the CNTs. This behavior is also observed in the damping results where the tan δ curves show the same trends. One difference however is that the magnitude of the tan δ peak in the second cycle, Fig. 5.6b, for the CNT composite is lower than that of the neat ABS. This result suggests, as observed also in Fig. 5.5, that the reinforcing effect leads to suppression of the dissipation mechanisms present in the matrix/rubber system. The constraint of both the butadiene particles and of the matrix surrounding the CNTs are possible explanations for the decrease in the observed damping.

The introduction of CNTs into the AM polymer shows an increase in the initial storage modulus of ~4.9%, Fig. 5.7a, suggesting that the CNTs act as reinforcing agents in the matrix. In addition, with the presence of the CNT, the energy dissipation at low strains (< 0.1%) is enhanced by the nanofiller. As the applied strain increases, the main dissipation mechanisms in the composite (rubber cavitation and crazing) take over and there is a sharp increase in damping. In the second cycle, at which point significant cavitation and crazing have occurred, the presence of the CNTs decrease the magnitude of the damping peak by ~7.1% as observed in Fig. 5.7b. This is attributed to the nanoscale interactions between the CNTs, polymer, and butadiene particles. There is possible CNT reinforcement at the butadiene/polymer interface, preventing the onset of cavitation, and thus crazing, in the material, as also observed in the injection molded samples in Fig. 5.6b. This leads to less material contributing to the overall damping of the sample. The reduction in storage modulus may suggest that the stick-slip mechanism is present, as observed in previous CNT composite studies [11, 12] where there is relative slip between the filler and matrix, however, it is not the case in this study since the decrease in storage modulus in the CNT composites occurs around the same strain and by the same magnitude as that of neat ABS. Therefore, no evidence of interfacial slip between the CNT and matrix is observed in these results. If this mechanism is present, it is overpowered and concealed by the dominant damping mechanisms in the matrix and butadiene particles, as previously described.

The effect of alignment is observed in the viscoelastic behavior as a function of temperature. As shown in Fig. 5.8b, AM induced polymer chain alignment restricts motion of the butadiene phase during its glass transition, evident in the lower loss

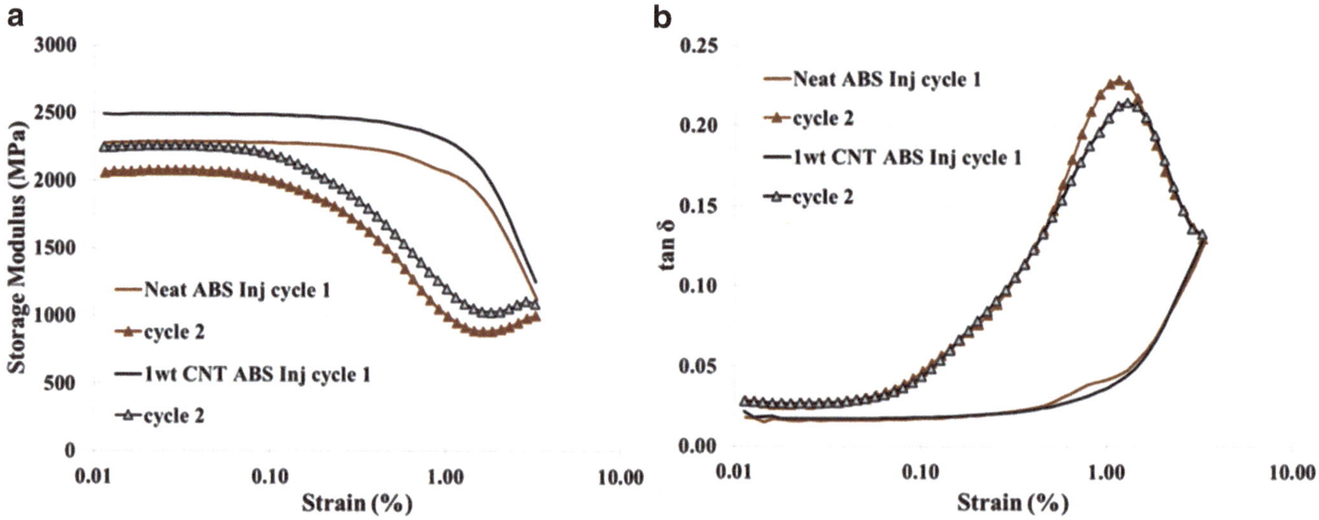

Fig. 5.6 (**a**) Storage modulus and (**b**) tan δ as a function of applied strain for neat ABS and 1 wt% CNT/ABS composites fabricated by injection molding (Inj)

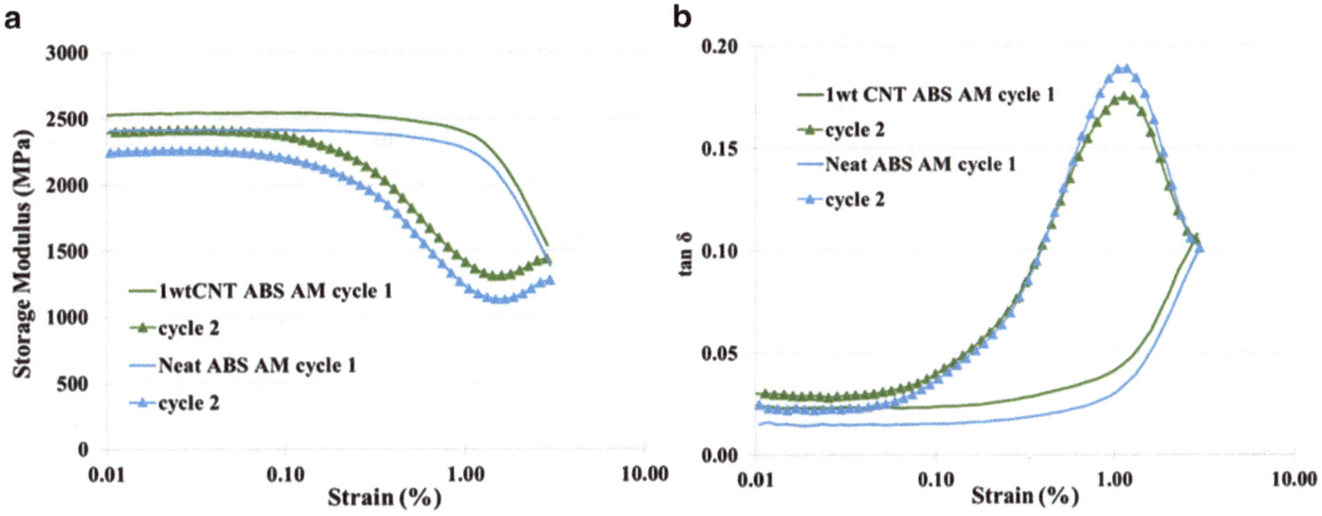

Fig. 5.7 (**a**) Storage modulus and (**b**) tan δ as a function of applied strain for neat ABS and 1 wt% CNT/ABS composites fabricated by additive manufacturing (AM)

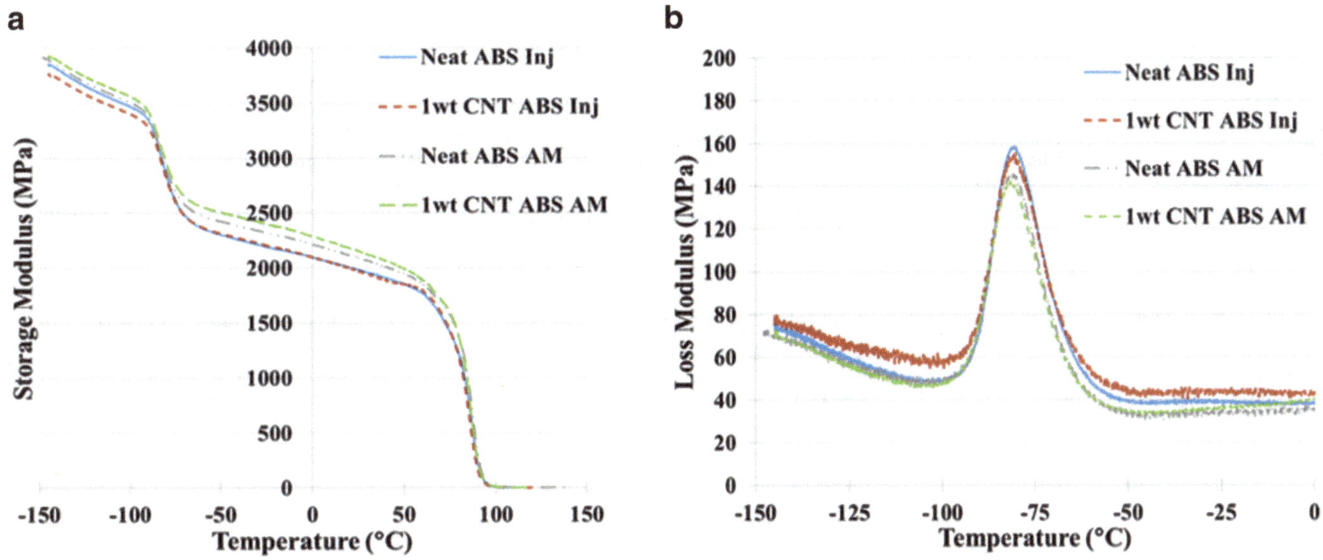

Fig. 5.8 (**a**) Storage modulus and (**b**) loss modulus as a function of applied temperature for neat ABS and 1 wt% CNT/ABS composites fabricated injection molding (Inj) and additive manufacturing (AM)

modulus peak. In addition, there is an increase in the storage modulus, Fig. 5.8a, after the butadiene transition between the neat ABS and CNT composite (both injection molded and AM) suggesting effective interaction between the butadiene particles and the CNTs, thus resulting in the reinforcement of the material.

5.4 Conclusion

A study on the effect of an additive manufacturing method on the internal microstructure of neat ABS and 1 wt% CNT/ABS composites was performed. It was found that the AM process induces significant polymer chain alignment in the printing direction (~25.7% programmed strain) which leads to an increase in the storage modulus of ~6.1% and a suppression of the rubber cavitation and crazing damping mechanisms evident in the tan δ peak decrease of ~17.4%. In addition, the presence of the CNTs leads to further reinforcement, evident as an increase in storage modulus of ~4.9% over AM neat ABS and further suppression of the damping mechanisms observed by a decrease of ~7.1% in the tan δ peak. This work sheds light on the changes the AM process has on both the microstructure and bulk dynamic behavior of both neat ABS and CNT/ABS nanocomposites. This study also provides initial insight into the development of multifunctional materials for dynamic applications. Future work will focus on further understanding of the effects of AM on the damping mechanisms of polymer nanocomposites and how to manipulate this behavior to obtain desired energy dissipation.

Acknowledgement FG is a contractor to the US Army Research Laboratory under Cooperative Agreement W911NF-16-2-0182.

References

1. Campbell, T., et al.: Could 3D printing change the world. In: Technologies, Potential, and Implications of Additive Manufacturing. Atlantic Council, Washington, DC (2011)
2. Tekinalp, H.L., et al.: Highly oriented carbon fiber–polymer composites via additive manufacturing. Compos. Sci. Technol. **105**, 144–150 (2014)
3. Ning, F., et al.: Additive manufacturing of carbon fiber reinforced thermoplastic composites using fused deposition modeling. Compos. Part B. **80**, 369–378 (2015)
4. Zhong, W., et al.: Short fiber reinforced composites for fused deposition modeling. Mater. Sci. Eng. A. **301**(2), 125–130 (2001)
5. Shofner, M.L., et al.: Nanofiber-reinforced polymers prepared by fused deposition modeling. J. Appl. Polym. Sci. **89**(11), 3081–3090 (2003)
6. Wahab, M. S., et al.: Development of polymer nanocomposites for rapid prototyping process. In: Proceedings of the World Congress on Engineering WCE, vol. 2. London (2009)
7. Postiglione, G., et al.: Conductive 3D microstructures by direct 3D printing of polymer/carbon nanotube nanocomposites via liquid deposition modeling. Compos. A: Appl. Sci. Manuf. **76**, 110–114 (2015)
8. Compton, B.G., Lewis, J.A.: 3D-printing of lightweight cellular composites. Adv. Mater. **26**(34), 5930–5935 (2014)
9. Crosby, A.J., Lee, J.-Y.: Polymer nanocomposites: the "nano" effect on mechanical properties. Polym. Rev. **47**(2), 217–229 (2007)
10. Moniruzzaman, M., Winey, K.I.: Polymer nanocomposites containing carbon nanotubes. Macromolecules. **39**(16), 5194–5205 (2006)
11. Suhr, J., Koratkar, N.A.: Energy dissipation in carbon nanotube composites: a review. J. Mater. Sci. **43**(13), 4370–4382 (2008)
12. Gardea, F., et al.: Energy dissipation due to interfacial slip in nanocomposites reinforced with aligned carbon nanotubes. ACS Appl. Mater. Interfaces. **7**(18), 9725–9735 (2015)
13. Ogasawara, T., Tsuda, T., Takeda, N.: Stress–strain behavior of multi-walled carbon nanotube/PEEK composites. Compos. Sci. Technol. **71** (2), 73–78 (2011)
14. Jar, P.-Y.B., et al.: Rubber particle cavitation on toughness enhancement of SMI-modified poly (acrylonitrile-butadiene-styrene). J. Polym. Sci. B Polym. Phys. **37**(14), 1739–1748 (1999)
15. Jar, P.-Y.B., Shinmura, T., Konishi, K.: A study of rubber particle cavitation in poly (acrylonitrile-butadiene-styrene)(ABS) using a freeze fracture technique. J. Mater. Sci. Lett. **19**(1), 73–75 (2000)
16. Ramaswamy, S., Lesser, A.J.: Microscopic damage and macroscopic yield in acrylonitrile–butadiene–styrene (ABS) resins tested under multiaxial stress states. Polymer. **43**(13), 3743–3752 (2002)

Chapter 6
Mechanical Properties of 3-D LENS and PBF Printed Stainless Steel 316L Prototypes

Wei-Yang Lu, Nancy Yang, Joshua Yee, and Kevin Connelly

Abstract Laser Engineered Net Shaping (LENS) and Powder Bed Fusion (PBF) are 3-D additive manufacturing (AM) processes. They are capable of printing metal parts with complex geometries and dimensions effectively. Studies have shown that AM processes create metals with distinctive microstructure features and material properties, which are highly dependent on the processing parameters. The mechanical properties of an AM material may appear to be similar to the corresponding wrought material in some way. This investigation focuses on the relationships among AM process, microstructure features, and material properties. The study involves several AM SS316L components made from 3D LENS and PBF printing. Specimens were taken from different locations and orientations of AM components to obtain the associated tensile properties, including yield, strength, and ductility, and to conduct microstructure analyses.

Keywords Additive manufacture • LENS • PBF • SS316L

6.1 Introduction

In a recent study of AM SS316L properties of a 3-D LENS deposited hexagon [1], tensile testing was performed for specimens oriented along the build direction (BD) and the transverse direction (TD). The results show that yield stress (YS) and ultimate tensile strength (UTS) of 3D LENS deposits are generally higher than nominal values for the wrought SS316L. As for ductility, LENS TD specimens were about the same as wrought SS316L; however, the elongation of BD specimens was notably low and varied greatly. Microstructure analyses suggest that the phenomena are caused by the inter-pass structure of the material. The BD specimen that had low strain to failure value was failed by inter-pass delamination, where lack-of-fusion defects are observed on the fracture surface of the specimen.

This work extends the investigation to SS316L prototypes of a different AM process PBF plus another LENS hexagon deposited from a new set of process parameters.

6.2 AM SS316L Prototypes

Figure 6.1a displays a substrate of prototypes 3D printed from a PBF machine ProX300 at Sandia National Laboratories/CA. Among these prototypes, a cylinder and a 3-tier hexagon, marked as 1 and 2 on the figure, were used for this study. The coordinate of each prototype was defined as the same as the machine reference frame: x-axis is the roller direction, y-axis is the transverse direction, and z-axis is the build direction. The diameter of the cylinder was 1.0 inch, and the height was 2.0 inch. For the hexagon, its cross-section (y-z plan) is shown in Fig. 6.1b. The dimensions were the same as the LENS hexagon, described in [1], where the distance between the outside edges of two opposite sides was 2.0 inch and the height was also 2.0 inch. Printing parameters and conditions are listed in Table 6.1.

A LENS fabricated short hexagon is shown in Fig. 6.1c. The dimensions were the same as the PBF hexagon described above but only had two tiers, where the total height was 1.2 inch instead of 2.0 inch. It was 3D printed from a LENS machine Optomec750 at UC Irvin. The hatch pattern was the same as that in [1], where the hatch lines of two adjacent layers were at 90°. They were parallel to either x or y axis. A new set of processing parameters is shown in Table 6.2.

W.-Y. Lu (✉) • N. Yang • J. Yee • K. Connelly
Sandia National Laboratories, Livermore, CA 94551-0969, USA
e-mail: wlu@sandia.gov

© The Society for Experimental Mechanics, Inc. 2018

J. Wang et al. (eds.), *Mechanics of Additive and Advanced Manufacturing, Volume 9*,
Conference Proceedings of the Society for Experimental Mechanics Series, DOI 10.1007/978-3-319-62834-9_6

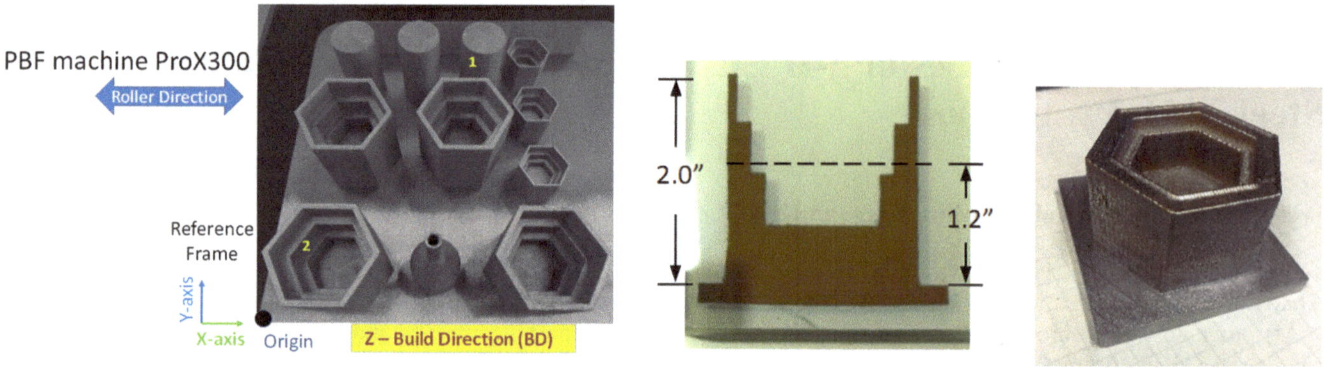

(a) (b) (c)

Fig. 6.1 AM SS316L prototypes. (**a**) 3D printed prototypes; (**b**) Hexagon cross-section in the y-z plan; (**c**) LENS short hexagon

Table 6.1 PBF processing parameters

Laser power	Power layer thickness	Laser scan speed
41% (of 500 W)	40 mm	1200 mm/s

Table 6.2 LENS short hexagon processing parameters

Laser power	Speed
400 W	16.3 mm/s

Fig. 6.2 Specimens machined from cylinders. (**a**) ID, location and orientation of each specimen; (**b**) Two specimen sizes

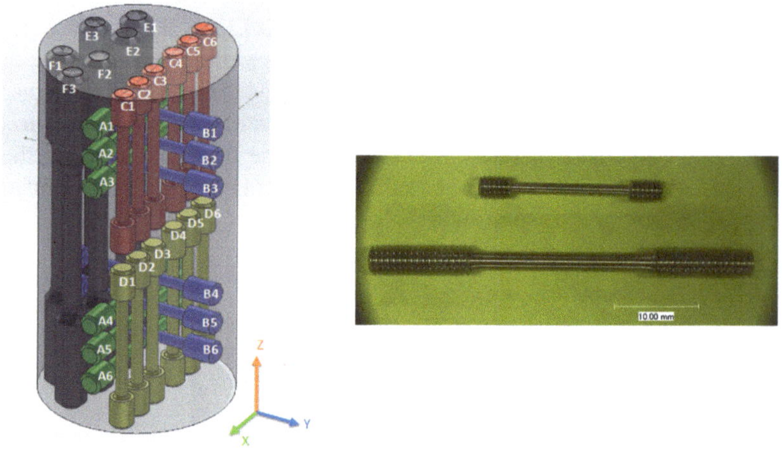

(a) (b)

6.3 Tensile Specimens

Tensile specimens were extracted from prototypes. Figure 6.2a shows the location, orientation, and size of each specimen machined from the PBF cylinder. These specimens could be gathered into five groups: Group A was specimens aligned in x-direction; Group B was in y-direction; Groups C, D and E were in z-direction. Again, x-, y-, and z-axes are referred to the machine reference frame. Two sizes of specimens were included, shown in Fig. 6.2b. The large size specimens, Group E, had a gage diameter of 0.1 inch with a total length of 2.0 inch. All others, Groups A – D, were the small size, which had the dimensions of 0.06 inch and 1.0 inch for gage diameter and total length, respectively. Furthermore, Group C and half of Groups A and B (A1 – A3 and B1 – B3) were from the top half of the cylinder, while Group D and other half of Groups A and B (A4 – A6 and B4 – B6) were from the bottom half of the cylinder.

Specimens in x-, y-, and z-directions were also machined from the hexagons. Due to the geometry, the specimens oriented in the build direction (z) were close to the wall; the transverse specimens, oriented in the x- and y-directions, were close to the base. Additional transverse off-axis specimens were also obtained. Their orientation x′ and y′ were 30° clockwise relative to x and y.

Fig. 6.3 Laser and mechanical extensometers are used to measure large and small deformations, respectively

6.4 Experimental Setup

Tensile tests were conducted on a MTS Bionix System, as shown in Fig. 6.3. The loading was quasi-static, at a strain rate of about 0.5×10^{-3} s^{-1}. The specimen was self-aligned with ball-and-socket joints in the loading train. Both laser and mechanical extensometers were used to measure the deformation at the gage section. The mechanical extensometer measures deformation up to 20% from a gage length of 0.3 inch. Sine it could not cover the whole deformation of SS316L specimens, the high resolution mechanical extensometer was used for small strain measurement only a laser extensometer was needed to capture the total deformation with a lower resolution. Thin reflection tapes were mounted on the gage section as targets for the laser extensometer, which were separated at least 0.45 inch.

6.5 Experimental Results

6.5.1 PBF Cylinder

The engineering stress-strain curves are plotted in Fig. 6.4a. There are 14 curves, about three specimens per each group except Group E with only two. The curves are clearly clustered in three bundles, which correspond to specimens in three orientations. All specimens along the build direction z, including Groups C, D and E, have a lower yield stress (YS) and ultimate strength (UTS); also they are relatively flat, indicating limited work hardening. Those specimens in the transverse directions, Groups A and B, have a similar UTS, but Group A, x-direction, shows a larger elongation while Group B, y-direction, has a higher hardening rate. The anisotropic behavior of the material is obvious.

In order to examine the data of z-direction specimens more closely, the hardening portion of the curves in Fig. 6.4a was magnified and shown in Fig. 6.4b. Two noticeable trends are displayed. One, the stress-strain curves of Group D specimens are all lower than Group C curves. Since Groups C and D correspond to top and bottom half of the cylinder, the result suggests that material properties may vary slightly along the build direction. The other is that Groups C and E curves are

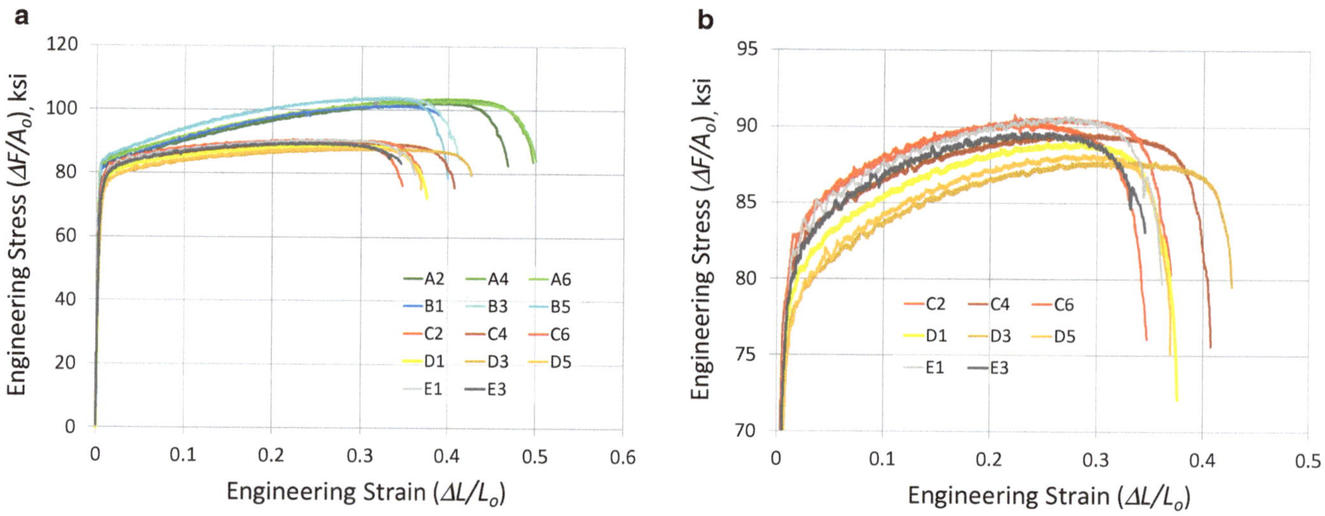

Fig. 6.4 Engineering stress-strain curves of specimens from the PBF cylinder. (**a**) A, B and C-D-E correspond to on-axis X, Y and Z curves; (**b**) Magnified hardening portion of Z curves

Table 6.3 Tensile parameters of PBF cylinder and wrought SS316L

Specimen	YS (ksi)	UTS (ksi)	Elongation (%)
A2	79	102	47
A4	80	103	50
A6	82	103	50
B1	80	101	39
B3	82	104	41
B5	81	104	40
C2	58	90	35
C4	74	90	41
C6	69	89	37
D1	63	89	38
D3	62	88	43
D5	55	88	37
E1	67	91	36
E3	68	90	35
Wrought [2]	**25**	**70**	**40**

generally overlap, which shows large and small size specimens have the same result. The small size specimen is adequate for AM material characterization.

The numerical values of these material parameters are listed on Table 6.3. Typical wrought SS316L properties are also included in the Table [2]. In comparison, YS and UTS of PBF printed SS316L are much higher than the wrought material. Although elongations are comparable, it is about 10–15% lower in the PBF build direction. Within a specimen group, the YS and UTS are quite consistent. The elongation, however, appears to have a relatively large scatter, especially within Groups C and D. Take Group C for example, the elongation of specimen C4 is far greater than C2 and C6. The result seems to correlate with the location of each specimen, that is C4 is closer to the center of the cylinder and C2 and C6 are closer to the edge. The result of Group D supports such correlation. The phenomenon may due to localized thermal transport and heat distribution [1], which vary between center and edge.

Obviously, more experimental data are needed to statistically confirm the location effects on AM materials that are suggested by the limited data presented in current experiments of PBF cylinder.

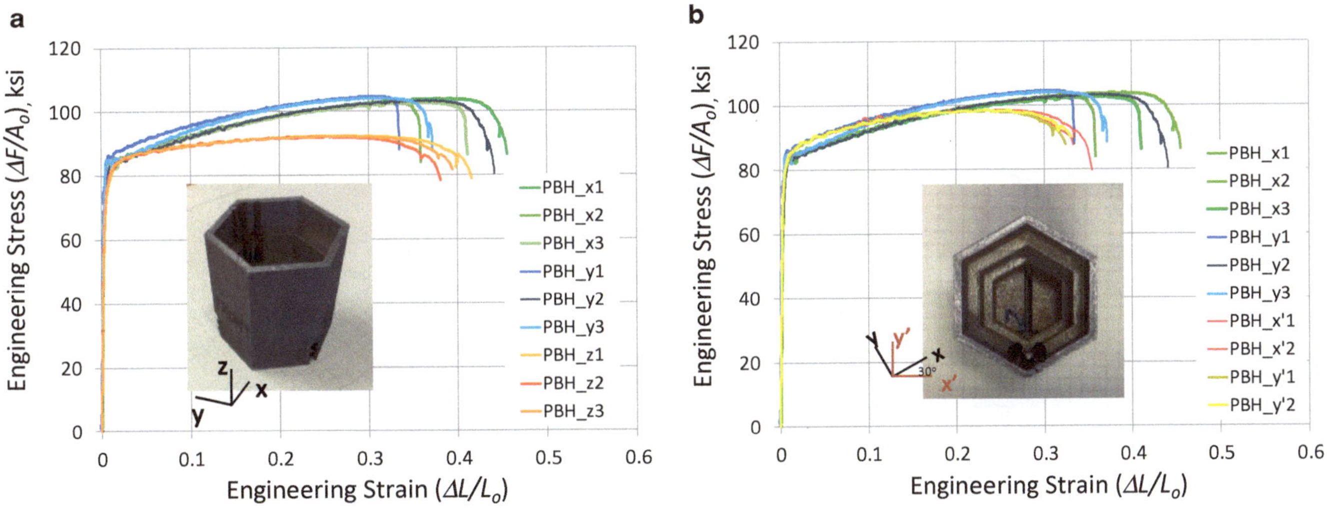

Fig. 6.5 Engineering stress-strain curves of specimens from the PBF hexagon. (**a**) X, Y and Z on-axis curves; (**b**) On-axis and off-axis X and Y curves

6.5.2 PBF Hexagon

The engineering stress-strain curves of on-axis specimens are plotted in Fig. 6.5a. Specimens in each orientation displayed distinctive hardening and elongation. These curves are consistent with those of PBF cylinder. Specimens of the same orientation show the same curve shape and material parameter values. The is one outlier, specimen PBH_y2, which is in the y-direction group but appears to follow the hardening behavior of those in the x-direction.

Figure 6.5b displays the engineering stress-strain curves of all transverse specimens of the PBF hexagon. The curves of two off-axis groups, x′ and y′, gather together exhibiting no difference. They are clearly distinctive from on-axis groups x and y, showing lower UTS and necking at a relatively low strain.

6.5.3 LENS Short Hexagon

The engineering stress-strain curves are shown in Fig. 6.6a and the material parameters are listed in Table 6.4. As shown in the figure, there are only two clusters for three groups of on-axis specimens. The engineering stress-strain curves of x and y group overlap displaying no difference between these two groups. It suggests that the material is transversely isotropic, but needs additional transverse off-axis data to confirm.

Comparing to the previous study [1], shown in Fig. 6.6b and Table 6.5, the current material is much improved, especially the properties in the build direction. The stress-strain curves are consistent and there is no premature failure, which indicates fabrication defects are controlled with the new set of processing parameters.

6.6 Fractography

The failure surfaces of PBF cylinder specimens were examined. Figure 6.7 shows a typical specimen from each group. At the low magnification 100×, voids are visible. The size and number of void do not appear to correlate with the strain to failure. As shown in Fig. 6.7d, specimen D3 displays a lot more large-size voids than C4 but the elongation is slightly longer. At the higher magnification 2000×, all failure surfaces exhibit typical ductile failure, featuring a small cell size dimpled fracture surface [1].

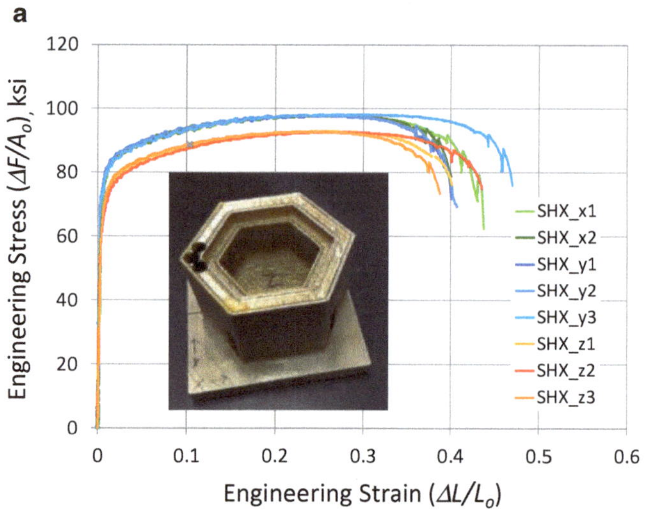

Short hexagon

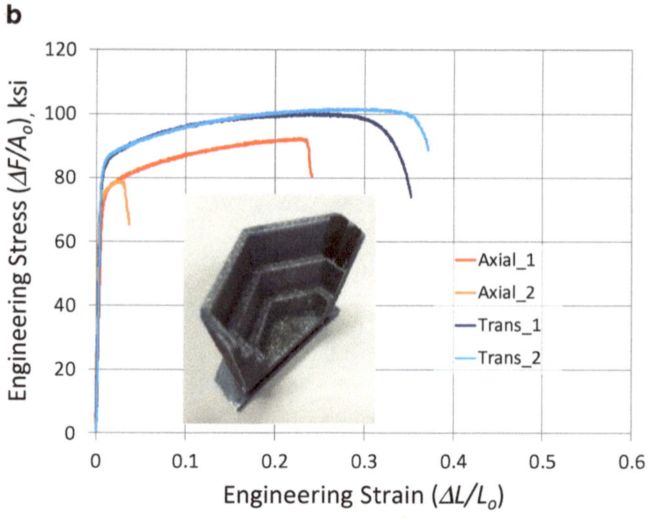

Hexagon [1]

Fig. 6.6 Engineering stress-strain curves of specimens from the LENS short hexagon. (**a**) Short hexagon; (**b**) Hexagon [1]

Table 6.4 Tensile parameters of LENS short hexagon

Specimen	YS (ksi)	UTS (ksi)	Elongation (%)
SHX_x1	70	98	44
SHX_x2	71	98	39
SHX_y1	72	98	40
SHX_y2	73	98	41
SHX_y3	71	98	48
SHX_z1	61	93	40
SHX_z2	60	93	43
SHX_z3	61	93	39

Table 6.5 Tensile parameters of LENS hexagon [1]

Specimen	YS (ksi)	UTS (ksi)	Elongation (%)
Axial_1	65	92	26
Axial_2	70	79	4
Trans_1	70	100	35
Trans_2	69	102	38

6.7 Summary and Conclusion

Tensile testing was conducted for specimens taken from PBF and LENS prototypes. Depending on the process, AM SS316L shows various anisotropic behaviors.

The PBF cylinder and hexagon included in this study were from the same substrate but relatively far apart. Even with different geometries, their bulk tensile properties are generally consistent.

Two different specimen sizes were considered. Their results are also consistent, which indicates using the small size specimen is adequate.

Within a PBF prototype, there are minor variations of mechanical properties with respect to location. This collaborates the suggestion that it is "a result of the localized thermal transport and heat distribution" [1].

3-D LENS and PBF prints exhibit cellular solidified cells with extremely fine arm spacing. Fractography of PBF cylinder sample shows the ductile dimple failure.

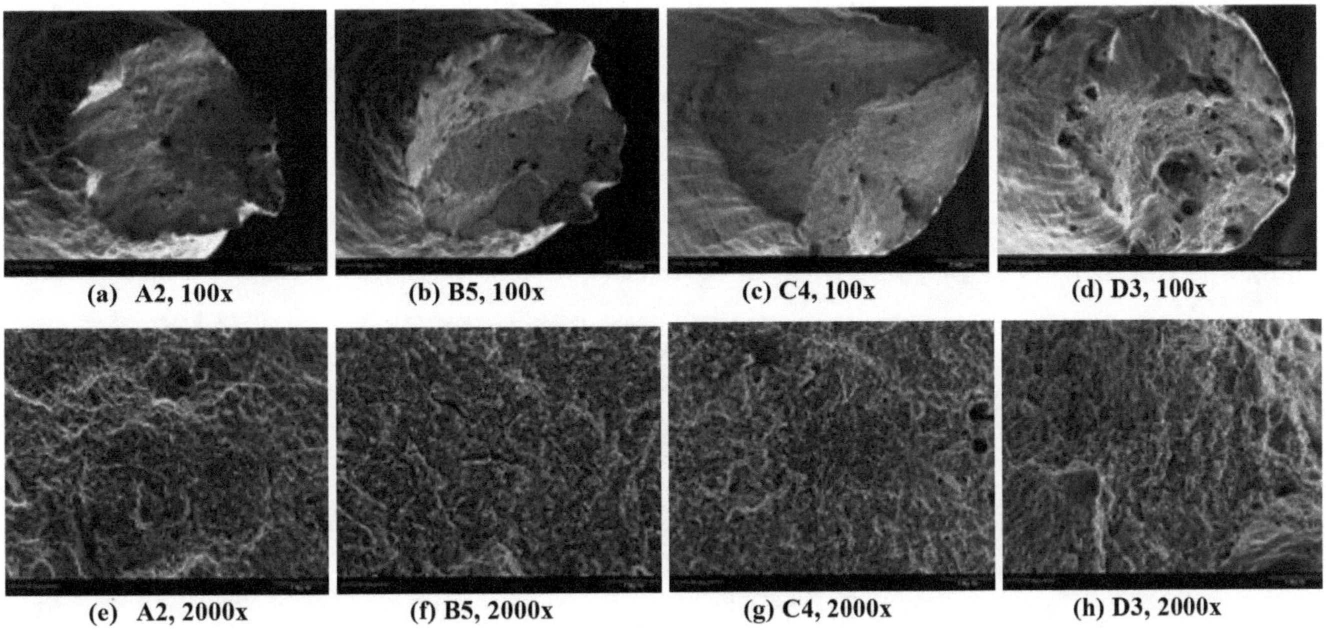

Fig. 6.7 SEM micrograph of failure surfaces. (**a**) A2, 100×; (**b**) B5, 100×; (**c**) C4, 100×; (**d**) D3, 100×; (**e**) A2, 2000×; (**f**) B5, 2000×; (**g**) C4, 2000×; (**h**) D3, 2000×

Microstructural analyses of failed specimens of PBF hexagon and LENS short hexagon are current underway. Also, tensile testing of LENS cylinder and 3-tier hexagon is being performed to compare with PBF data.

Acknowledgements The authors would like to acknowledge the LENS hexagon provided by University of California, Irvine. Sandia National Laboratories is a multi-mission laboratory managed and operated by Sandia Corporation, a wholly owned subsidiary of Lockheed Martin Corporation, for the U.S. Department of Energy's National Nuclear Security Administration under contract DE-AC04-94AL85000.

References

1. Yang, N., et al.: Process-structure-property relationships for 316L stainless steel fabricated by additive manufacturing and its implication for component engineering. J. Therm. Spray Technol. (2016). doi:10.1007/s11666-016-0480-y
2. www.AZOM.com

Chapter 7
Effect of Heat Treatment on Friction Stir Welded Dissimilar Titanium Alloys

Kapil Gangwar and M. Ramulu

Abstract A significant drop in the hardness values have been achieved in the friction stir welded dissimilar titanium alloys after post weld heat treatment. A closer look at the microstructural evolution suggests that the heat treatment at 933 °C changes the microstructure in the weld nugget from globular blocky grains to distinct microstructure with β flakes distributed uniformly in the α phase. With Ti-6242 FG (β transus = 996 °C), and Ti-54 M (β transus = 966 °C) being on the retreating and advancing side respectively, a series of microstructure evolves from retreating to advancing side. With a difference in the composition of the alloys, difference in the diffusion of elements is also responsible for the corresponding microstructure. Diffraction patterns were measured by X-ray diffraction method. Diffraction peaks for ADV side were slightly shifted towards left in comparison with retreating, and center of the weld along with significant difference in the preferred crystal orientation between as welded and post weld heat-treated conditions.

Keywords Friction stir welding • Dissimilar titanium alloys • Microstructure • Microhardness • SEM/EDS • XRD analysis

7.1 Introduction

Friction stir welding (FSW), needing no longer an introduction towards friction based additive manufacturing processes, has been studied extensively for softer and as well as harder materials [1–3]. FSW, by virtue of producing a solid-state joint poses lesser threats to structural integrity in comparison to fusion welding. A great number of studies focusing on the effect of welding parameters on microstructural, and mechanical properties evolution in titanium alloys has been dedicated in the literature. Furthermore, material flow, temperature histories, phase transformation, texture evolution, residual stress have also been studied in detail for various titanium alloys [4]. Wu et al. have extended our understanding towards achieving super-plasticity in FSWed Ti-6Al-4 V at various temperatures demonstrating the variation in the thickness of α lamella and aspect ratio [5]. Drawing on the conclusions of Wu et al., our study builds on the literature adding towards the FSW of dissimilar titanium alloys. Despite its invention in 1991, and applications since, literature on dissimilar titanium alloys is scanty [6]. To the best of our knowledge, this is the first study of its kind demonstrating the microstructural evolution, hardness variation, and elements' migration in FSWed dissimilar titanium alloys for AW and PWHTed conditions.

FSW parameters, if changed slightly, have a tremendous impact on the evolving temperatures and corresponding microstructure in the weld nugget [7]. The complexity of the FSW processes is further convoluted when two dissimilar titanium alloys with different composition and different β transus temperatures are concerned [8]. Concerning with the structures made of different titanium alloys, operating under different loading and temperature conditions, it is of great significance to study the underlying characteristics of welded joints. Therefore, the objective of this study is (i) to understand the microstructural evolution in AW and PWHTed conditions; and (ii) to correlate the microstructural changes with emerging mechanical properties along with the diffusional characteristics of the alloying elements.

To our knowledge, this study is the first to quantify microstructural disparities in FSWed dissimilar titanium alloys. In light of novelty of these findings, we have elucidated the effect of heat treatment on the microstructure, microhardness, elements migration, and X-ray diffraction patterns.

K. Gangwar (✉) • M. Ramulu
Department of Mechanical Engineering, University of Washington, Seattle, WA 98195, USA
e-mail: kapildg@uw.edu; ramulum@u.washington.edu

© The Society for Experimental Mechanics, Inc. 2018

J. Wang et al. (eds.), *Mechanics of Additive and Advanced Manufacturing, Volume 9*,
Conference Proceedings of the Society for Experimental Mechanics Series, DOI 10.1007/978-3-319-62834-9_7

7.2 Experiments and Procedure

As milled sheets of distentions 200 × 100 × 4 mm^3 Ti-6242 FG and Ti-54 M were FSWed at a rotation speed of 275 rpm and a transverse speed of 100 mm min^{-1}. A W-La alloy welding tool with an undisclosed geometry has been used with proper thermal management (via water cooling) of the tapered tool, and the backing anvil [9]. Argon shielding was used to prevent the surface oxidization and excessive flash formation during welding. Before FSW, the sheets were cleaned to avoid oxidation of titanium. Advancing (ADV) and retreating (RET) side consist of Ti-6242 FG, and Ti-54 M, respectively in this study. Microstructures and microhardness profiles were observed on the transverse cross section of the welded specimens. Electron dispersive spectroscopic analysis has been carried out to study the elements migration in, and around the weld nugget. X-Ray diffraction patterns were observed at the center of the transverse cross section (1.5 mm below the top surface) of the weld at three different locations-ADV, center (CEN), and RET. Post weld heat treatment (PWHT) was carried at 933 °C for 45 min. The specimen was furnace cooled at a rate of 15 °C until room temperature.

7.3 Results and Discussion

Figure 7.1 shows the typical appearance of the macrostructures as observed in both condition. As observed in the macrostructures, no visible significant change in the morphology of the weld has been observed. However, mixing is pronounced and a significant amount of material from RET is being transported in the weld nugget. In the top of the weld, major amount of material is coming from the RET side, while in the center of the weld the microstructure is primarily dominated by Ti-54 M. In the bottom of the weld, due to incomplete penetration of the tool, a distinct boundary between two alloys is visible.

Figure 7.2 is the magnified image of figure one demonstrating the clear difference in the width of crescent like microstructure as we move from RET to ADV side. Noteworthy is the point that the effect of heat treatment conditions have no significant effect on the morphology of the weld for the major element in considerations is titanium. Alloying elements such as Sn, and Al (with lower melting point than 933), and Zr, V, and Mo (with higher melting point than 933 °C)

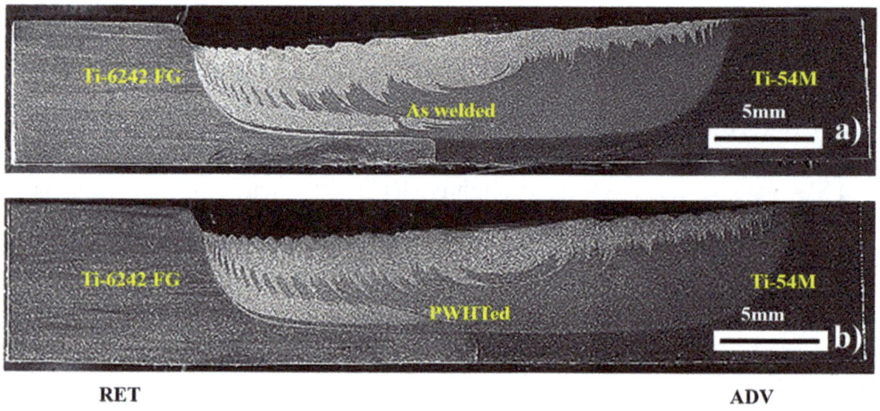

Fig. 7.1 Typical appearance of the macrostructure for as-welded and PWHTed FSWed dissimilar titanium alloys

Fig. 7.2 Morphology of transverse weld under adopted processing condition

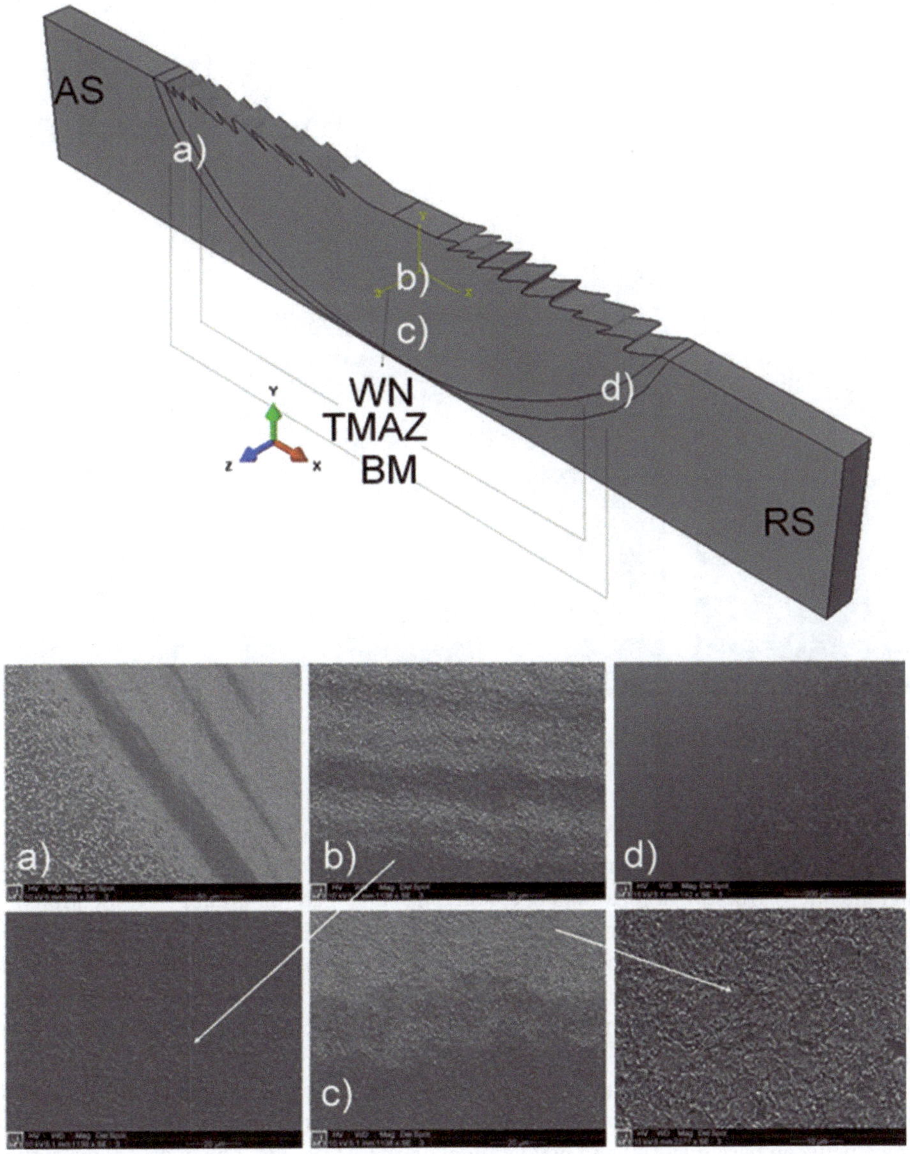

Fig. 7.3 Prescribed location for microstructural observation, and corresponding microstructure. *White* region is the region which is lighter in appearance and consist of Ti-6242 FG. Darker region is dark in appearance and consists of Ti-5 4M

have significant effect on the evolving microstructure as is discussed later. A line of defect in the bottom of the weld has been observed due to lack of penetration.

Figures 7.3 and 7.4 represent the underlying microstructure of at prescribed locations on the transverse cross section of the weld. Considering Figs. 7.3 and 7.4 together it can be observed that the base material (Ti-6242 FG, location a) microstructure has changed from coarser globules β to rather finer β at the expanse of $\beta \rightarrow \beta + \alpha$ transformation.

Figure 7.3b represent the microstructures observed in the center of the weld. Magnified microstructure of the darker and lighter regions is presented, and indicated by the white arrows. As can be seen that the microstructure in case of Fig. 7.3b for darker phase consist of equiaxed refined β indicating that the material undergoes intense mechanical stirring but heat cycle above β transus temperature. An interesting point to observe is the fact that the darker phase which is being carried away from Ti-54 M (β transus = 966°C) located on the RET side has undergone above β transus temperature. However, as observed in Fig. 7.2, where center of the weld consist of material from the RET side, experiences a rather uniform cooling pattern in comparison to top, and, bottom of the weld. Thermal gradients, as they develop, are steeper on the top and bottom of the weld due to maximum exposed areas of the weld. As a result, the center of the weld, the darker phase, primarily coming from RET side (in our study Ti-54 M), results in formation of refined equiaxed β grains. Consequently, the top of the weld, primarily consisting of Ti-6242 FG (also above (β transus = 990°C) results in formation of lamellar microstructure

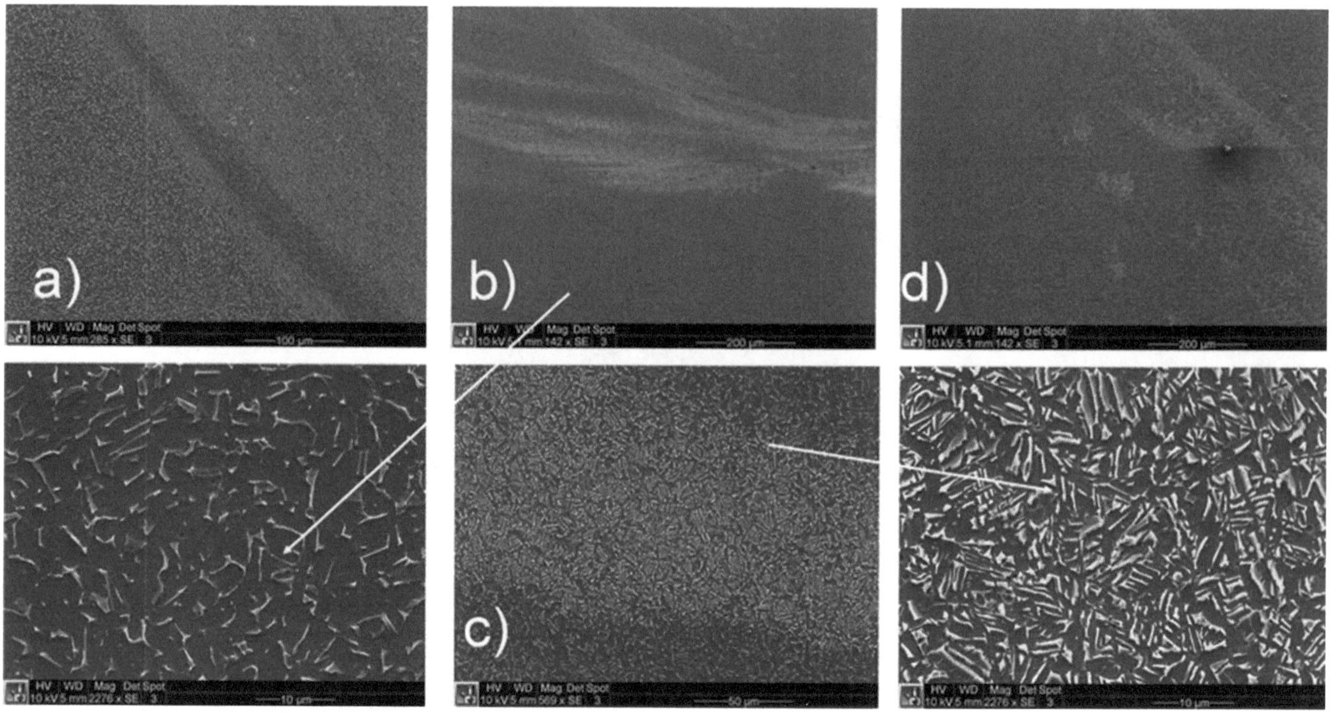

Fig. 7.4 Prescribed location for microstructural observation are similar to Fig. 7.2, a magnified image of *black* and *white* regions is indicated by *white arrows*

due to faster cooling rate, as observed in the lighter region of Fig. 7.3b. In Fig. 7.3d, the microstructure has evolved into refined equiaxed α, along with bimodal microstructure.

Upon heat treating, the microstructure at locations b, c, and d has changed remarkably. At location b, the darker region in the center of the weld, the microstructure transformed from $\beta \rightarrow \beta + \alpha$. PWHT Time and temperature for this darker region, primarily consisting of Ti-54 M, were enough to transform the recurring β into coarser equiaxed α and transformed β phases. The microstructure of white phase, however, consisting of Ti-6242FG, transforms into coarser lamella of α. Since, the PWHT was significantly lower than the β transus temperature of Ti-6242FG, complete transformation of β phase (as observed in Fig. 7.3b), did not occur.

Figure 7.5 depicts the profile of microhardness on the center of transverse cross section of the weld at a depth 1.5 mm below the top surface. The microhardness as observed on the ADV side of the base material is similar for AW, and PWHT condition. The interface between base material and weld nugget on the ADV side, however, shows higher values of hardness in comparison with PWHTed condition. The banded microstructure in case of AW and PWHTed condition consist of consist of lamellar, and finer equiaxed grains respectively. These results comply with the fact that microstructure with finer lamella results in higher values of hardness. As we approach the center of the weld the hardness values begin to drop in both cases. Since, the center of the weld primarily consists of material from RET side (in our study Ti-54 M), it has been observed that the untransformed α transforms into recrystallized β grains resulting in lower hardness values. The interface on the RET side, unlike banded appearance on the ADV side, is quite distinct. Nonetheless, the presence of TMAZ could not observed in the SEM micrographs. The increment in the microhardness values at the interface is a result of refined β grains. The tapered geometry of the tool corresponds with higher temperature, and higher thermal gradient on the top of the specimen. The bottom of the specimen also experiences higher temperature due to tool transverse and rotation.

Considering four different locations as shown in the schematic of WN in Fig. 7.6 following can be added. Microstructure in region 1, and 2 consists of material on the ADV side. Due to higher temperature on the ADV side a bit coarser lamella in region 1, in comparison with region 2, are observed. Region 3, which is in the center of the weld, and is primarily consist of material migrating from the RET side, refined β grains, due to lesser thermal gradient in the center, are observed. Region 4, consisting material from the ADV side, in between top and bottom produces a range of microstructure depending on the lamella width [10]. Microhardness contours on the transverse cross section of the weld are presented in the Fig. 7.7. A distinct boundary between the weld nugget and parent material is visible in both cases.

A distinction between ADV (Ti-6242FG) and weld nugget is lucid in AW, and PWHTed condition. Indicating that the adapted PWHTed conditions have a little effect on material with higher β transus temperature (Ti-6242FG). Nonetheless, a

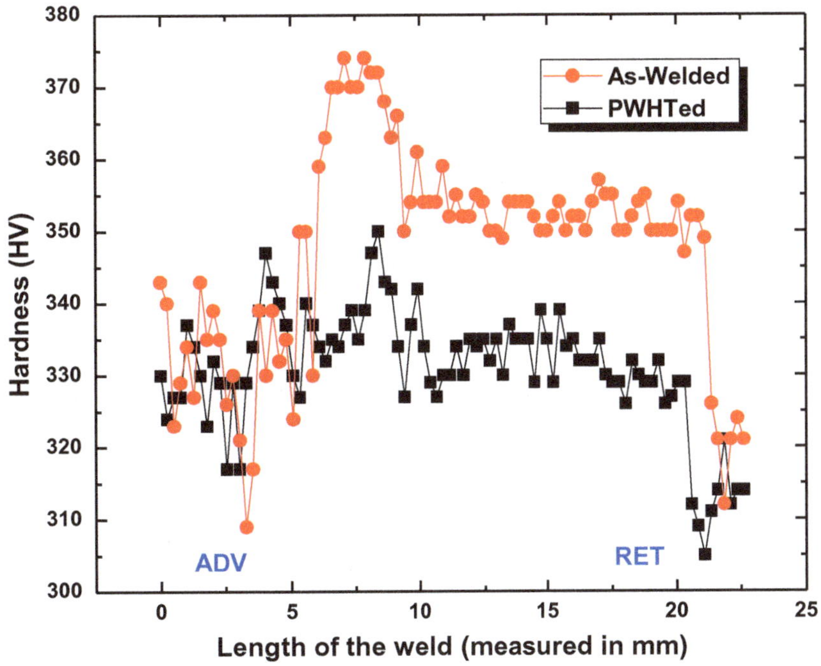

Fig. 7.5 Microhardness profile as observed in the center of the cross section of transverse weld (at depth 1.5 mm from top)

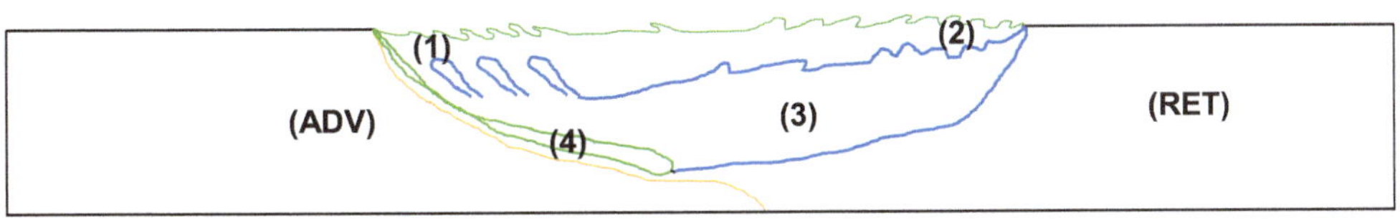

Fig. 7.6 Distinct regions in weld nugget for the case of friction stir welded dissimilar titanium alloys

significant hardness drop, throughout the center of the weld, has been observed. Which could be attributed to the fact that alloying element vanadium (V), migrating from retreating side. Vanadium not only stabilizes the β phase but also the subsequent $\beta \rightarrow \beta + \alpha$ transformation, due to higher temperatures on the ADV side experiences a significant broadening in the α lamella resulting in the lower hardness values. Contours on the RET side, however, are discombobulated as far as the boundary between Ti-54 M and WN is concerned. Grain refinement is the primary reason for higher hardness values on the RET side boundary in case of PWHTed conditions.

Electron dispersive spectroscopic maps for AW, and PWHTed conditions are shown in the Figs. 7.8 and 7.9. In Fig. 7.8, the EDS maps are presented at the center of the weld in AW condition. While Al is the second common constituent in both alloys, and the annealing temperature is 933 °C, no distinct boundary between darker and lighter phase has been observed for aluminum. Ti, V, Zr, and Mo show distinct boundary between darker and lighter phase affirming the migration of elements in regions 1–4 as shown in the Fig. 7.6. Sn, despite lower melting temperature than annealing temperature, shows distinct region between two phases due to the fact that Sn is only constituent of Ti-6242FG, and not of Ti-54 M. Also, the contribution of Sn, and Zr (coming from Ti-6Al-2Sn-4Zr-2Mo) is lower. While Zr and Sn provide solid solution strengthening, effect of such mechanism remains out of scope of this article. In increment in Sn (7–15 at.%) result in decreasing elastic modulus and plasticity of titanium alloys [11].

The microstructural evolution throughout the weld (from top to bottom, and from ADV to RET) is clearly a function of thermal gradient, and the presence of α *or* β stablizers. Looking into the two different regions of darker phase that appears in Fig. 7.8, the microstructure in the vicinity of the center if characterized by the lamella microstructure, while in the vicinity of bottom the microstructure consist of refined equiaxed β grains having insufficient time to promote $\beta \rightarrow \beta + \alpha$ transformation.

Diffraction patterns for ADV, CEN, and RET side are presented in the Fig. 7.10. For the purpose of clarity, the diffraction patterns are shifted with respect to each other along vertical axis. The ADV side, which is primarily Ti-6242 FG, a near α alloy consist of mainly α phase. In both cases, the diffraction pattern shows that the phase composition was $\alpha + \beta$ phase. For PWHTed condition, the texture of all phases is relatively stronger in comparison with AW condition. Furthermore, effect of alloying composition is visible in terms of relative peak shift in comparison to pure α titanium. Which could be attributed to

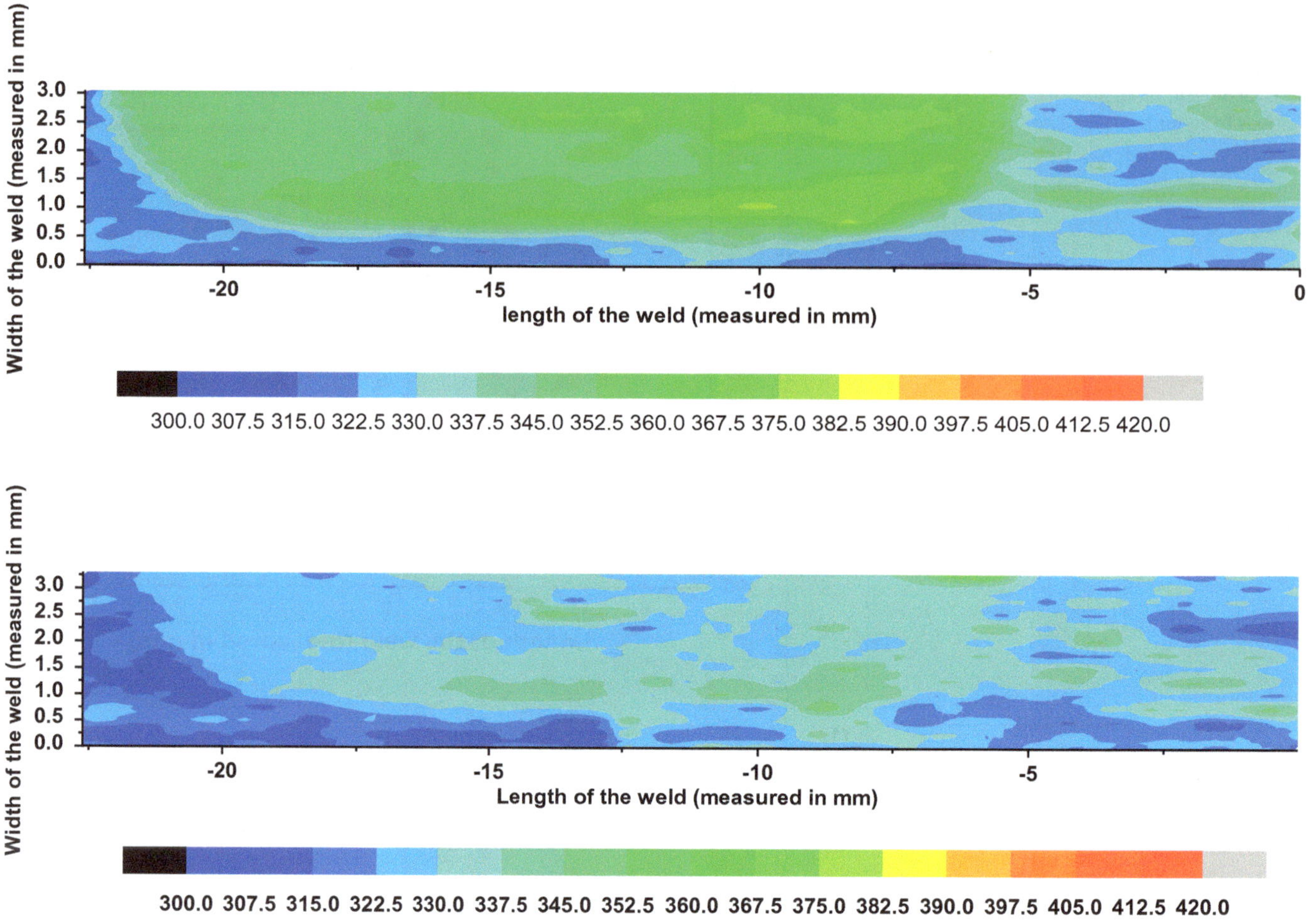

Fig. 7.7 Microhardness contours on the transverse cross section of the weld

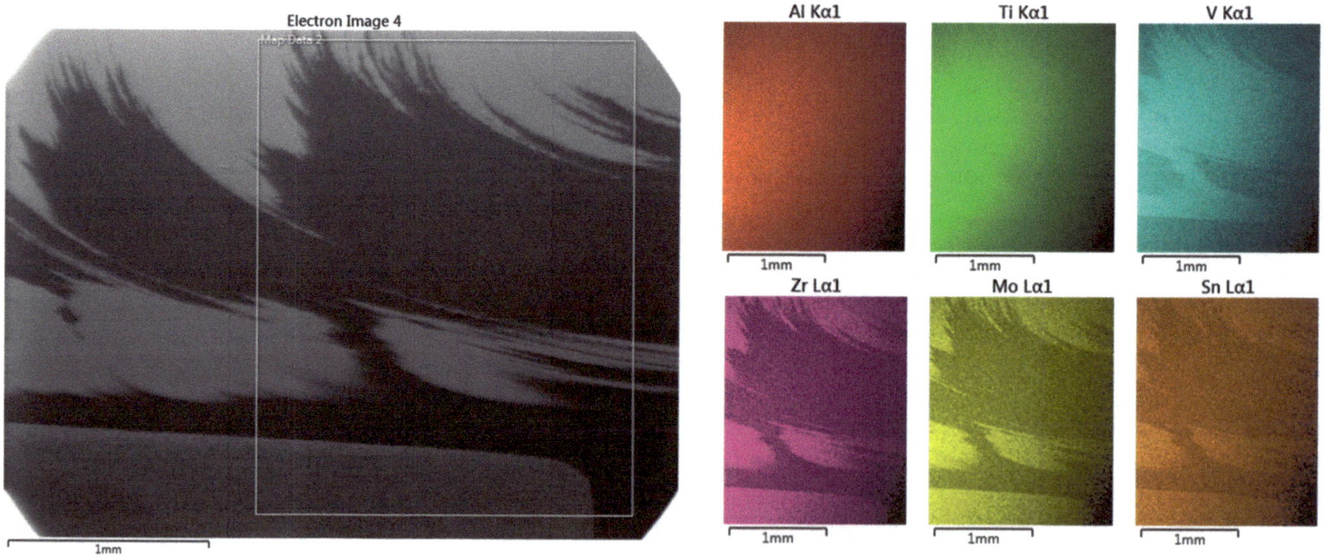

Fig. 7.8 EDS map at the center of the weld for AW condition

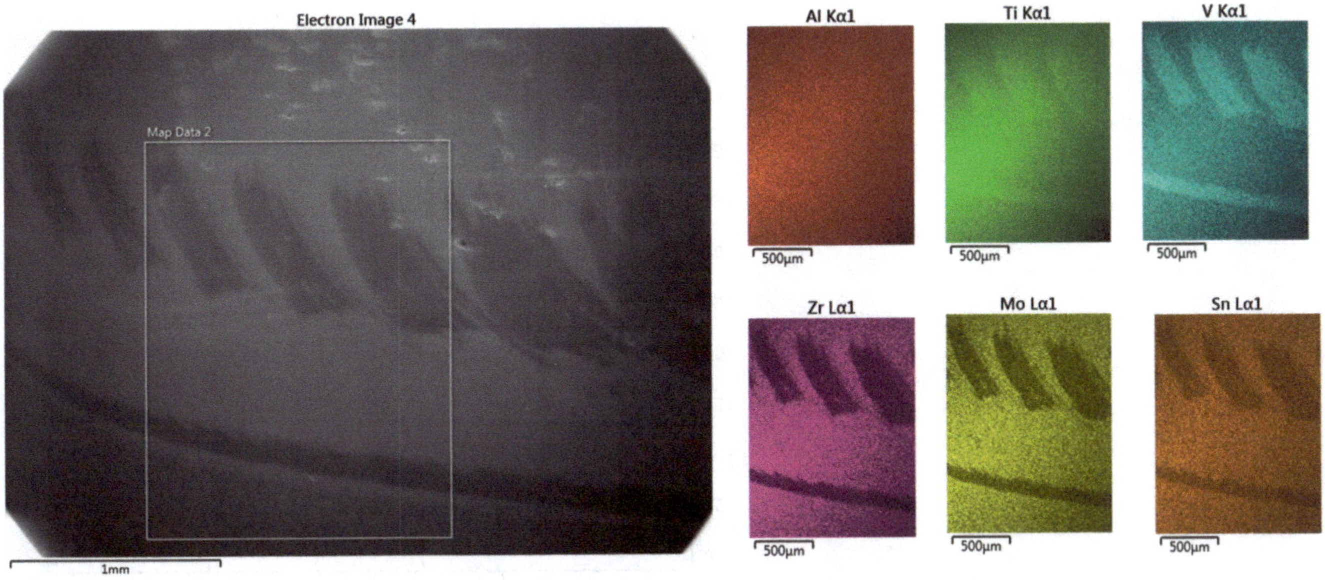

Fig. 7.9 EDS maps at the ADV boundary for PWHTed condition

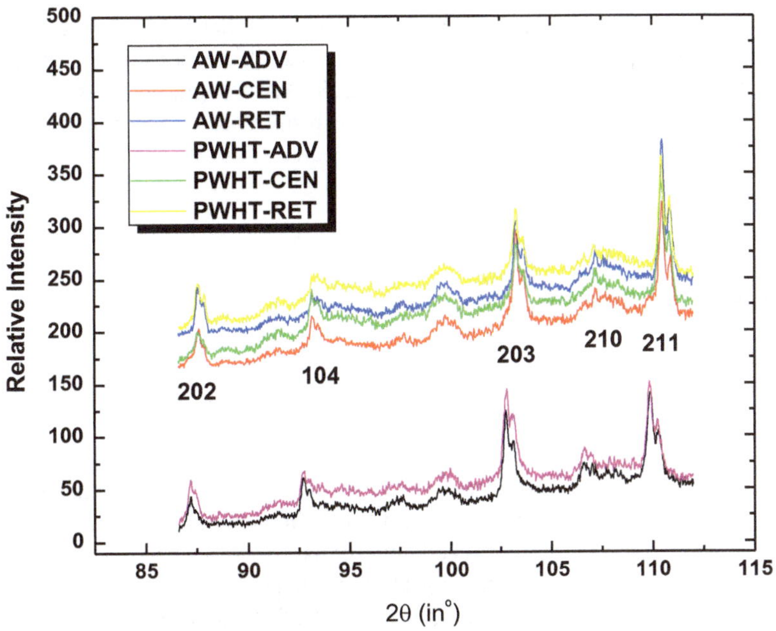

Fig. 7.10 X-Ray diffraction patterns at three points located in the center of the transverse cross section of the weld, at depth 1.5 mm

the change in *c*/*a* ratio of α *hcp*. A detailed study by electron back scattered diffraction, or synchrotron XRD, or neutron diffraction is required to obtain subsurface information about the crystal orientation. Furthermore, the diffraction patterns as observed for CEN and RET side of the weld are nearly identical (peaks are shifted vertically for clear description). On the RET side and in the CEN of the weld, the microstructure consist of $\alpha + \beta$ lamella. As opposed to lamella microstructure as observed on the ADV side, the corresponding cooling rate, and emerging thermal gradient governs the growth, and aspect ratio of α lamella. Variants of α transforms, and grows during $\beta \rightarrow \beta + \alpha$ transformation. Associated thermal gradient throughout the weld nugget from top to bottom gives rise to different amounts of $\beta + \alpha$ at different stages. The evolving microstructure, or lamella morphology at these stages corresponds with different levels of residual stresses. In addition, the alloying composition with α , *and* β stabilizers, and neutral elements the change in *c*/*a* ratio is also accountable for the peak shift in the diffraction patterns [12].

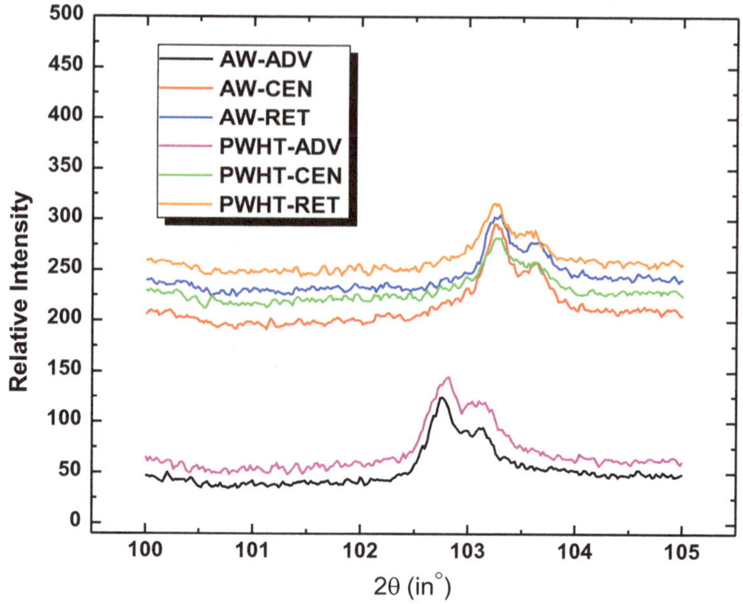

Fig. 7.11 Full width half maximum (FWHM) pattern for AW, and PWHTed conditions. The peaks are shifted vertically relative to each other

Upon looking at Fig. 7.11, it can be observed that the sub-peaks (green) of β have nearly disappeared for the PWHTed condition in the center of the weld. Again, this difference can be noted in terms of refined α lamella in case of AW condition, unlike coarser lamella in PWHTed condition. Although, out of scope of this article, it can be observed that the FWHM also increases in PWHTed conditions. These results are in accordance with the fact the residual stresses, and corresponding dislocation density, and microhardness decrease for the PWHTed condition.

7.4 Conclusions

In this article we have presented the effect of PWHTed conditions (933 °C for 45 min, and furnace cooling at rate of 15 °C/min until room temperature) on FSWed dissimilar titanium alloys (near α {Ti-6Al-2Sn-4Zr-2Mo}, and $\alpha + \beta$ {Ti-54 M}) processed at 275 rpm and 100 mm min^{-1}. Following conclusions can be drawn.

1. Primary microstructures on ADV, and RET side consist of banded, and continuous morphology respectively.
2. Micro-hardness values drop significantly in the weld nugget. Microhardness of material with higher β transus temperature is unaffected by the PWHTed conditions.
3. EDS analysis showed the migration of elements from RET side to ADV side. Except Al, all the elements ascertained a distinct boundary with the other elements. Solid solution strengthening analysis as an effect of Sn has not been attempted in this study.
4. XRD patterns, observed in the center of the transverse cross section of the weld (1.5 mm below top surface) revealed presence of α *and* β phases on ADV, CEN and RET side. A rather stronger texture of corresponding phases has been observed in PWHTed condition. The shift in the peaks between ADV, and CEN & RET has been observed. This phenomenon could be attributed to the elements' migration (corresponding with changes in the *c*/*a* ratio), and to the changes in microstructure from finer to coarser lamella.

Acknowledgements We sincerely thank Dr. Dan Sanders, and The Boeing Company for providing the material and for continuous support. We express our gratitude toward University of Washington for supporting our research.

References

1. Sanders, D.G., Edwards, P., Cantrell, A.M., Gangwar, K., Ramulu, M.: Friction stir-welded titanium alloy Ti-6Al-4V: microstructure, mechanical and fracture properties. JOM. **67**(5), 1054–1063 (2015)
2. Hong, J.K., et al.: Friction stir welding of dissimilar Al 5052 to Ti-6Al-4V alloy with WC-Co tool. Steel Res. Int. **81**(9), 1092–1095 (2010)
3. Fazel-Najafabadi, M., Kashani-Bozorg, S.F., Zarei-Hanzaki, A.: Joining of CP-Ti to 304 stainless steel using friction stir welding technique. Mater. Des. **31**(10), 4800–4807 (2010)
4. Edwards, P.D., Ramulu, M.: Material flow during friction stir welding of Ti-6Al-4V. J. Mater. Process. Technol. **218**, 107–115 (2015)
5. Wu, L.H., Xiao, B.L., Ni, D.R., Ma, Z.Y., Li, X.H., Fu, M.J., Zeng, Y.S.: Achieving superior superplasticity from lamellar microstructure of a nugget in a friction-stir-welded Ti-6Al-4V joint. Scr. Mater. **98**, 44–47 (2015)
6. Gangwar, K., Ramulu, M., Cantrell, A., Sanders, D.G.: Microstructure and mechanical properties of friction stir welded dissimilar titanium alloys: TIMET-54M and ATI-425. Metals. **6**(10), 252 (2016)
7. Zhang, Y., Sato, Y.S., Kokawa, H., Park, S.H.C., Hirano, S.: Microstructural characteristics and mechanical properties of Ti-6Al-4V friction stir welds. Mater. Sci. Eng. **485**(1–2), 448–455 (2008)
8. Leyens, C., Peters M.: Titanium and Titanium Alloys: Fundamentals and Applications. Wiley-VCH; John Wiley distributor, Weinheim Chichester. p. xix, 513 pages (2003)
9. Edwards, P., Ramulu, M.: Surface residual stresses in Ti-6Al-4V friction stir welds: pre- and post-thermal stress relief. J. Mater. Eng. Perform. **24**(9), 3263–3270 (2015)
10. Ji, S.D., Li, Z.W., Wang, Y., Ma, L.: Joint formation and mechanical properties of back heating assisted friction stir welded Ti-6Al-4V alloy. Mater. Des. **113**, 37–46 (2017)
11. Gorna, I.D., Bulanova, M.V., Valuiska, K.O., Bega, M.D., Koval, O.Y., Kotko, A.V., Evich, Y.I., Firstov, S.O.: Alloys of the Ti-Si-Sn system (titanium corner): phase equilibria, structure, and mechanical properties. Powder Metall. Met. Ceram. **50**(7–8), 452–461 (2011)
12. Malinov, S., Sha, W., Guo, Z., Tang, C.C., Long, A.E.: Synchrotron X-ray diffraction study of the phase transformations in titanium alloys. Mater. Charact. **48**(4), 279–295 (2002)

Chapter 8
Effect of Porosity on Thermal Performance of Plastic Injection Molds Based on Experimental and Numerically Derived Material Properties

Suchana A. Jahan, Tong Wu, Yi Zhang, Jing Zhang, Andres Tovar, and Hazim El-Mounayri

Abstract Plastic injection molding industry uses traditionally machined tools and dies to manufacture various sizes and shapes of plastic products. With the advent of advanced manufacturing technology and expanding global competition in business, it is necessary to provide innovative solutions to the injection molding industry to sustain their business. Typically, the cooling time comprises more than half of the overall injection molding cycle time. The application of additive manufacturing technique can provide a solution to reduce the cooling time in injection molding process. The potential of 3D printing technology to produce any size and shape of products using metal powders provides an opportunity to design and produce innovative injection molding tools, which is unattainable by traditional machining process. Though the conformal cooling channels are capable of reducing the cooling time significantly, the cost of manufacturing the injection molds by 3D printing is quite high and hence a crucial decision making factor for the mold designers about whether or not to go for the 3D printed molds. By making the molds porous, it is possible to reduce the cost of additive manufacturing, thus creating a positive impact on the use of 3D printed molds in injection molding business. In this paper, the effect of mold porosity on the thermal performance of the injection molds are studied. The properties of 3D printed mold material and traditional mold material is quite different and have been considered for the analysis. An optimization study has been conducted to identify the best possible design solution in terms of thermal and printing cost perspectives.

Keywords Injection molding • Additive manufacturing • Conformal cooling • Porous mold • Optimization

8.1 Introduction

In the present world, a wide variety of plastic products are used in everyday life. Injection molding is a major part of the plastic industry, consuming a large percentage of the total amount of plastics [1]. Plastic injection molding is a multipurpose process to obtain different complex sizes and shapes of high quality products from thermoplastic and thermosetting materials with the application of heat and pressure [2]. To obtain a better-quality plastic part the design of the injection molding tooling, specifically the design of die core and cavity is very critical. It also plays an important role in the economic aspects of the business. The cooling of injection molding tooling plays very important role in the total production cycle time of the injection molding process. Time is significant in the entire molding process, as it constitutes about half of the time in the overall production cycle [3]. Figure 8.1 shows a generic distribution of time in the total injection molding process. Injection molding is a widely used and accepted manufacturing process for the production of plastic parts. Traditionally straight holes are drilled into the solid dies to cool the hot molten plastic inside the cavity. This cooling process takes a major portion of the production cycle, leading to high cost of production. At present, with the rising competition worldwide in the plastic product business, it has become very important to lower the production cost, one way of doing so is to reduce the production cycle time. With the currently available manufacturing methods of traditional dies, it is not possible to reach much closer to the cavity wall; precisely, to the rounded corners or sharp edges of an intricate geometry. Hence an innovative technological solution is essential to solve the problem, to cope up with the ongoing business competition. Using conformal cooling channels is a good option for this purpose. Conformal cooling channels are the ones, which "conform" to the shape of the internal cavity wall, thus can reach much closer to the molten plastic material, compared to the cases of traditional molds. With the use of advanced additive manufacturing technology, the shape and size of the channels are no longer limited to larger circular cross-section, rather it is up to the requirement of the optimal channel configuration.

S.A. Jahan (✉) • T. Wu • Y. Zhang • J. Zhang • A. Tovar • H. El-Mounayri
Department of Mechanical Engineering, Purdue School of Engineering and Technology,
Indiana University – Purdue University Indianapolis, 723 West Michigan Street, SL 260, Indianapolis, IN 46202, USA
e-mail: sjahan@iupui.edu

© The Society for Experimental Mechanics, Inc. 2018

J. Wang et al. (eds.), *Mechanics of Additive and Advanced Manufacturing, Volume 9*,
Conference Proceedings of the Society for Experimental Mechanics Series, DOI 10.1007/978-3-319-62834-9_8

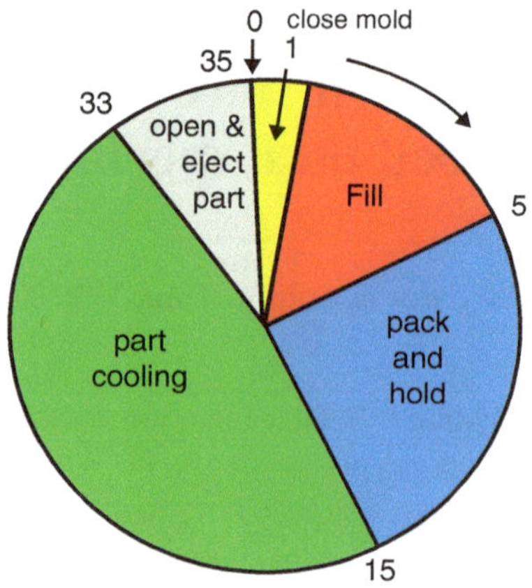

Fig. 8.1 Typical cycle time in injection molding [3]

The use of cooling channels conformal to the molding cavity improves the control of mold temperature and part dimensions. This has been reported by a group at MIT in the 1990's. This group studied the design and production of tooling with conformal channels for quite a long time [4–8]. With the advent of this technology, a number of studies have been carried out about the numerical and experimental analysis of cooling channels till then. Different profiles and geometric creation of cooling channels have been a significant area of analysis in this field [9–23]. In recent years, detailed study of the design parameters of conformal cooling channels and thermo-mechanical analysis has been carried out [24–27]. A recent study analyses the effect of porosity of the thermal properties of porous 3D printed materials [28]. Hence it is important to identify the thermal performance of 3D printed molds, which is the focus of this study.

8.2 Methodology

8.2.1 Design of Conformal Cooling Channels

Conformal cooling channel are potential alternatives of traditional straight cooling channels in plastic injection mold cavities and cores. Typically, straight holes are drilled into the cavity of the molding tool to create a passage of cooling water to flow inside. But due to the limitation of traditional manufacturing processes, it is difficult, sometimes impossible to create the cooling channels closer to the mold or cavity walls. As a result, the cooling time in the total production cycle of injection molding becomes significantly long. On the contrary, the conformal cooling channels 'conform' to the shape and profile of the plastic part, or the profile of the mold wall. With the application of additive manufacturing technology, any size and shape of cooling channel can be created inside the mold, however closer to the wall the mold designer may need. Thus, the conformal cooling channels can significantly reduce the cooling time of plastic injection molding production cycle. The most common design for conformal cooling channels in the mold cavity is the spiral design. This means, a spiral configuration of cooling channel goes into the mold body, conforms to the outside profile of the cavity wall, and then gets out; creating a continuous passage of coolant to flow. In traditional injection molds, the core has less space (compared to mold cavity) inside for a spiral design of cooling channels. In such cases, a simple U-shape configuration is a good design option for conformal cooling channels in mold core.

In Fig. 8.2, a simplified generic mold design with conformal cooling channels is shown. The mold is used to manufacture a simple cylindrical plastic part (shown in green). It is 2 mm thick plastic part, with 40 mm height and 40 mm outer diameter. The mold has two basic parts- the core, and the cavity. The core has U-shaped conformal cooling channel and the cavity includes a spiral design conformal channel. The channels have 10 mm diameter circular cross-section in both core and cavity. The spiral channel has pitch distance 20 mm. The channels in cavity are placed 20 mm from the mold wall, whereas they are 10 mm apart from mold wall in the core. The cavity and core blocks are 100 × 100 × 70 mm in size. All the measurements and design shapes are influenced from previous research and publications of the authors, the additive manufacturing research group at Purdue School of Engineering and Technology, IUPUI.

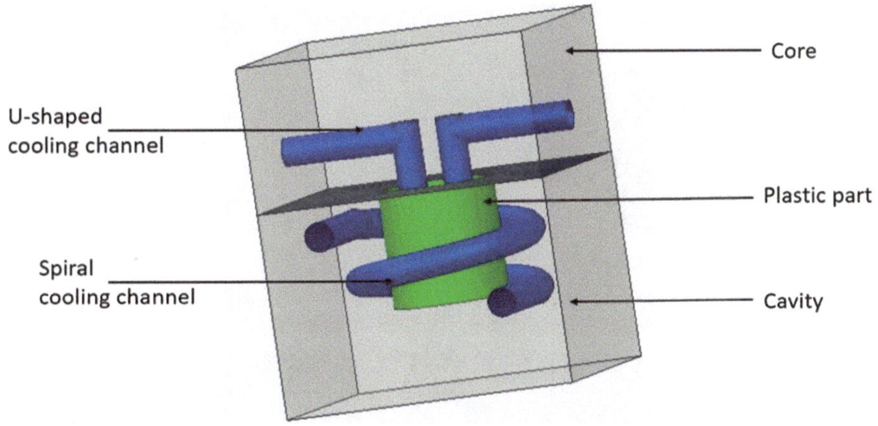

Fig. 8.2 Schematic simplified generic injection mold with conformal cooling channels

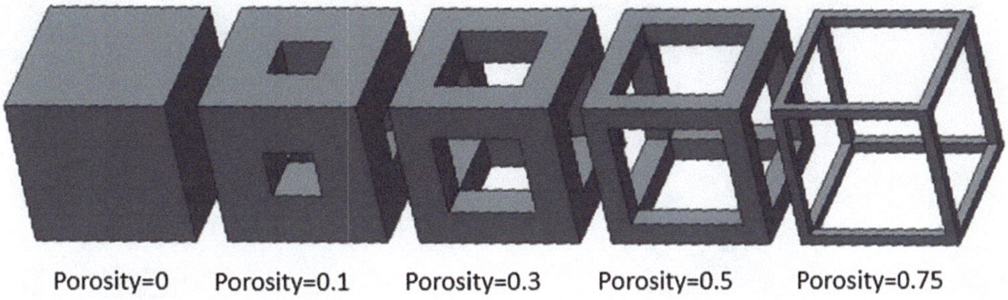

Fig. 8.3 Schematic of unit cells with different porosities

8.2.2 Design of Porous Injection Mold

The cost of production in additive manufacturing is quite high at present, especially for metal printing. The more metal powder is needed, the more is the cost. Hence, a porous structure of the 3D printed product is a good solution in this problem. If a well-designed porous cavity and core can be printed, which simultaneously ensures the thermal performance of the mold, with structural stability, this would provide the mold designers with a potential source of improved business. To identify the most effective porous design of the mold, four different porosities along with the zero porosity, i.e.; the solid design are considered. The porosities are 0.10, 0.30, 0.50 and 0.75. Fig. 8.3 shows CAD models of unit cells of each kind of different porosities. Five different molds with these five different porosities are analyzed for the thermal performance in this paper.

8.2.3 CAD Model

The solid mold is the same as shown in Fig. 8.2. The other molds have specific porosity incorporated into them. The modeling in done using Creo Parametric software. Each side is broken down into unit cells with specific porosities. Due to the presence of cooling channels, unit cells closer to those features, are kept fully solid. Due to geometric limitations, mold cavity and cores with 0.50 and 0.75 porosities has less number of porous unit cells compared to the 0.10 and 0.30 porosities. Figure 8.4 shows the CAD models of the porous molds.

8.2.4 Thermal Analysis Model

During the injection molding process, hot molten plastic is injected into the mold cavity via the injection unit and cooling water is passed through the cooling channels. The water serves two purposes. First, to cool down the hot plastic and second, to warm up the cavity and core body. This warming up process is required to ensure minimum shrinkage of the plastic part

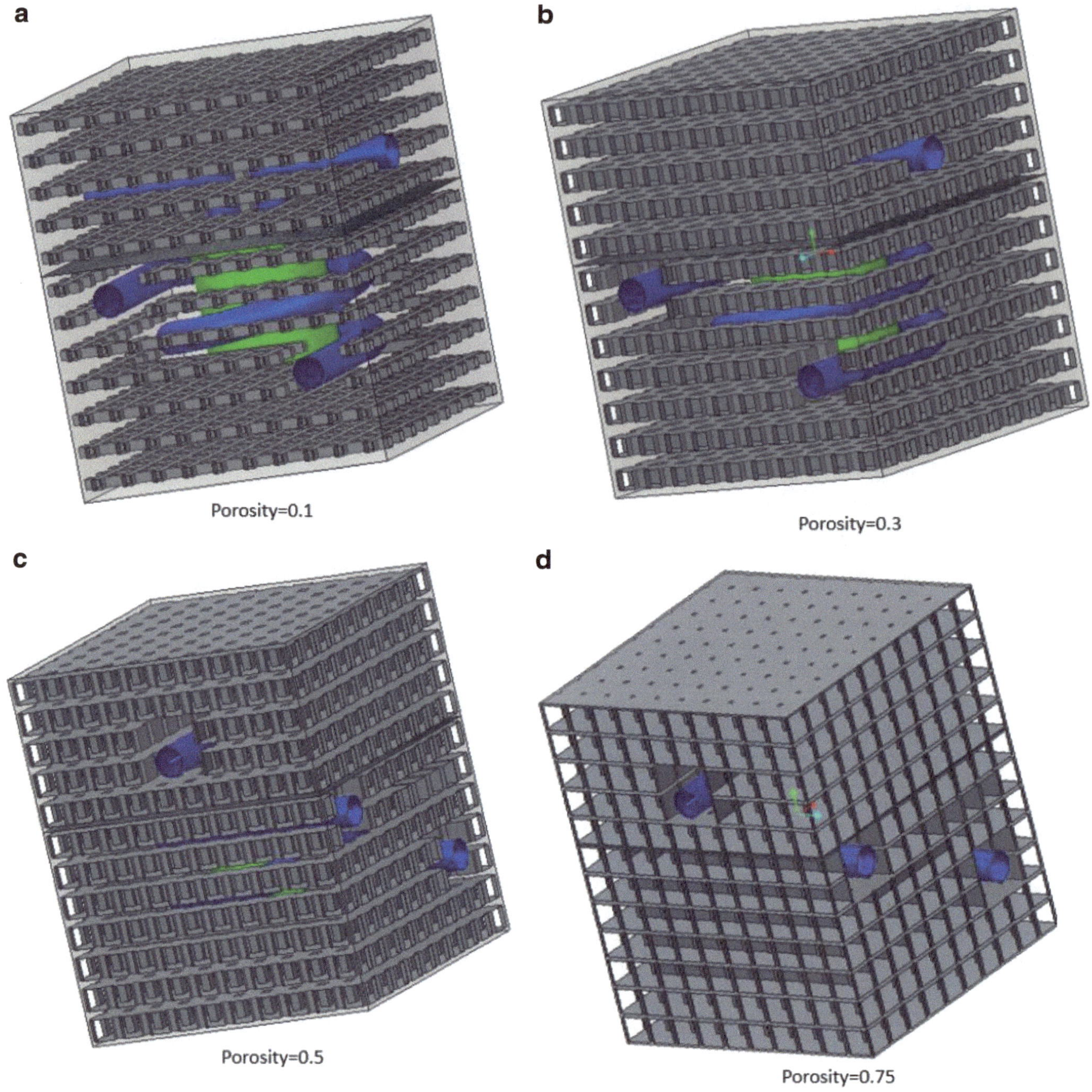

Fig. 8.4 Full CAD model of molds (**a**) porosity 0.1, (**b**) porosity 0.3, (**c**) porosity 0.5, (**d**) porosity 0.75

being produced. Hence this is understandable that the cooling water's temperature is needed to be above the temperature of the mold cavity and core. The water temperature is kept couple of degrees higher than room temperature.

As the study is concerned on the cooling process and the phenomena happening after the molten hot plastic is inside the cavity; at time, t = 0, cooling water starts running inside the channel. The molten plastic is considered to be already inside the cavity.

To analyze the thermal behavior of the mold with hot molten plastic inside the cavity and cooling water running through the channels, a transient thermal analysis model is developed using ANSYS™ workbench 14.5. An initial temperature of the molten plastic of 168 °C, and a water temperature at the inlet of 28 °C were used. Convection heat transfer co-efficient is applied on the cooling channel surfaces. This convection heat transfer co-efficient is based on the flow rate of water inside the cooling channels. Fine meshing with medium smoothing is applied on the geometry using automatic mesh generator in ANSYS™.

Table 8.1 Material properties of plastic and experimentally driven properties of SOLID stainless steel

Material	Thermal conductivity (W/m-K)	Density (kg/m^3)	Specific heat (J/kg-K)
Polypropylene	0.14	830	1900
Solid stainless steel	13.8	7800	460

Table 8.2 Properties of porous steel

Porosity	Thermal conductivity (W/m-K)	Density (kg/m^3)
0	13.1	7800
0.1	10.22	7020
0.3	5.71	5460
0.5	3.19	3900
0.75	2.54	1950

The plastic material is polypropylene and the mold material is stainless steel. The material properties of the polypropylene plastic are obtained from material supplier's datasheet or material specification documents.

It is here important to note that the material properties of a traditional structural steel block are not the same as the material properties of a 3D printed stainless steel block. As the current research concentrates on 3D printed molds, hence the specific properties of the 3D printed stainless steel is required. The metal powder used in the study is 15-5 PH1 Stainless steel powder. The properties are derived experimentally using traditional material specimen testing methods. The properties of 15-5 PH1 is published in EOS data sheet [29]. These properties are based on solid steel block (porosity = 0). On the contrary, a finite element analysis (FEA) study has been conducted in recent times to obtain the thermal properties of porous 3D printed 15-5 PH1 stainless steel specimen [28]. This study provides homogenized thermal conductivity values of porous stainless steel. These two different arrays of material properties have been considered to analyze the thermal performance of the mold in this study.

The solid (porosity = 0) mold has been analyzed using ANSYS™ with different thermal conductivity values at different porosities. In each case, different cooling time is obtained. On the other hand, the different porous CAD models are analyzed with solid material properties, which provides another set of cooling time results. These two sets are compared and analyzed to obtain a trend of the effect of porosity on thermal performance of injection molds. Tables 8.1 and 8.2 shows the material properties used in the analyses:

8.3 Results and Discussion

8.3.1 Thermal Analysis Results

The transient thermal analysis provides results as the cooling time required to cool down the hot molten plastic to 50 °C. This is a reference temperature typically used in industry. At this temperature, the plastic part is ejected from the molding machine which is the end of the ongoing production cycle and after that, another new cycle starts. At this ejection temperature, the plastic part is hotter than the mold body. The temperature of the mold body varies between 35 and 45 °C at this point.

As mentioned earlier, the thermal analysis was divided into two sets. First set with the properties derived from FEA study, and applied on solid mold CAD model. The second set of analysis is conducted with experimentally derived material properties which is applied on the Porous CAD models. Figures 8.5 and 8.6 shows the temperature distribution on the plastic part at the ejection time. Figure 8.5 is for the FEA derived properties and Fig. 8.6 is for experimentally derived properties.

In each of the figure, the cooling time for each case is marked with a red box for easy understanding. In the first set, where the numerically driven properties are applied on the fully solid model, the cooling time varies from 42.91 s for porosity = 0–77.99 s for porosity = 0.75. On the other hand, where only the property of solid material property is applied on manually created porous model, the cooling time varies from 39.96 s to 49.85 s respectively. This variation in two sets can be attributed by two factors.

First, considering porosity = 0, the CAD model is exactly same, i.e.; the fully solid model. In the first set, the thermal conductivity of mold material is 13.1 W/m-K, whereas in the second case, the thermal conductivity is 13.8 W/m-K. Hence there is about 5% discrepancy between the material property derived in different methods. This contributes largely to the

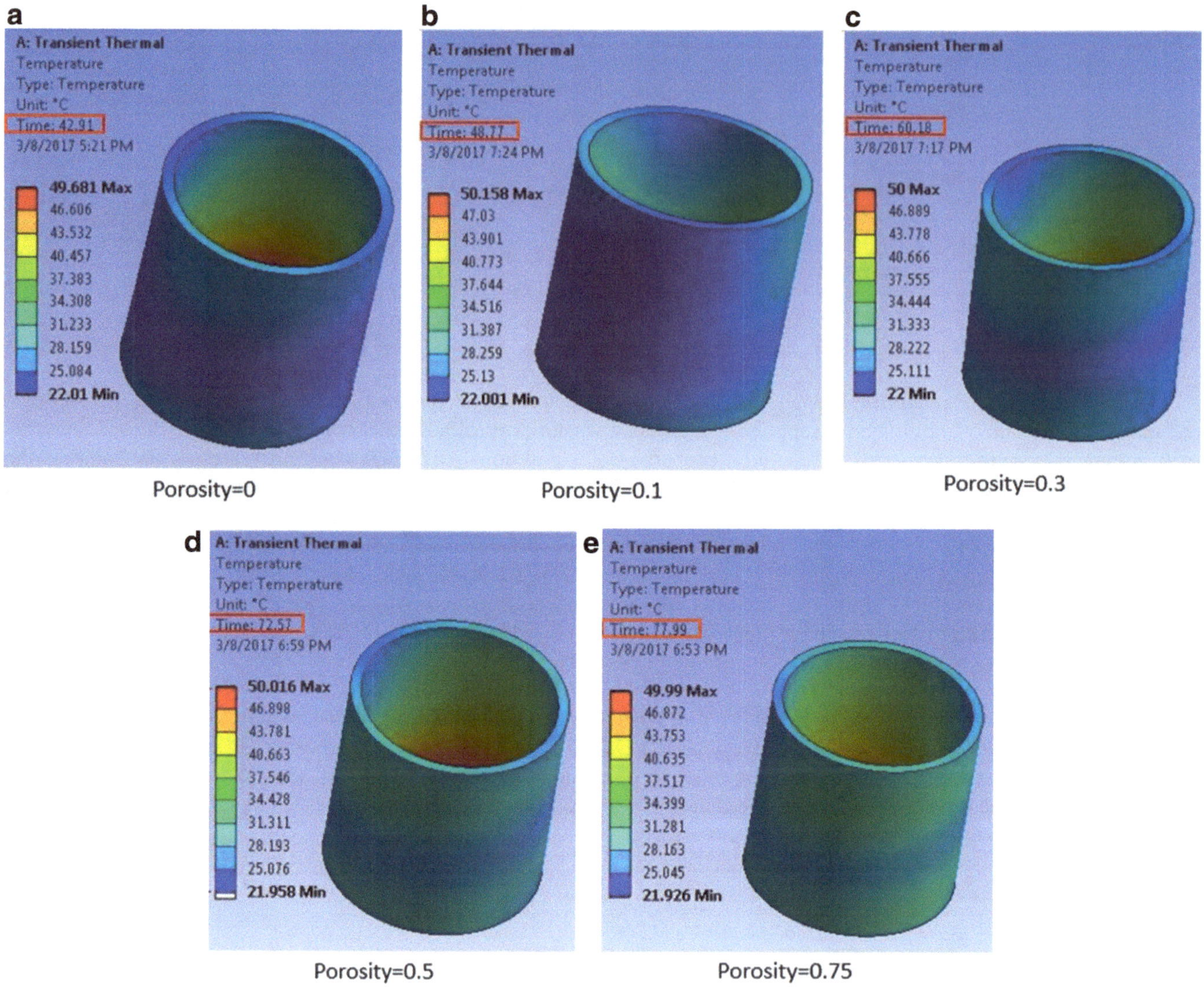

Fig. 8.5 Temperature distribution on plastic part at ejection, analysis with FEA derived properties (**a**) porosity 0, (**b**) porosity 0.1, (**c**) porosity 0.3, (**d**) porosity 0.5, (**e**) porosity 0.75

cooling time variation between two sets of cases. With higher value of thermal conductivity, the cooling time decreases, as expected.

Second, in the first set of models, the mold cavity and core is considered as uniformly porous, and uniform property has been applied on the whole body of core and cavity. On the other hand, in the second set, the porous structure is created by manual modeling in drawing software. This process considers the presence of cooling channels and other features of the mold, where uniform porosity is not viable, to ensure the continuous and undisturbed flow of coolant inside the channels. As a result, excess fully solid area with higher value of thermal conductivity contributes to the lower cooling time value. Hence, this geometric discrepancy is another important factor towards the variation of cooling times between two sets of results.

A trend analysis is done about the effect of porosity on thermal performance, as shown in Fig. 8.7. Here, dimensionless cooling times are taken here to understand the trend better. As we can see, for the numerically derived properties, the dimensionless cooling time varies from 1 to 1.81, whereas for experimentally derived properties, the variation is 1–1.25. From this analysis, the trendline equation for both the cases can also be derived. These equations can provide an estimation of cooling time at any other porosities that has not been considered in the current study. For the FEA derived properties, dimensionless cooling time can be calculated as $y = -0.8398x^2 + 1.7478x + 0.9845$ and for experimentally derived properties: $y = -0.2948x^2 + 0.5114x + 1.025$.

In the cost analysis of 3D printed products, it is simply considered that the larger or heavier a product is, the higher is the production price. Till date, the per unit cost of metal powder is quite high that creates a negative impact of the effectiveness

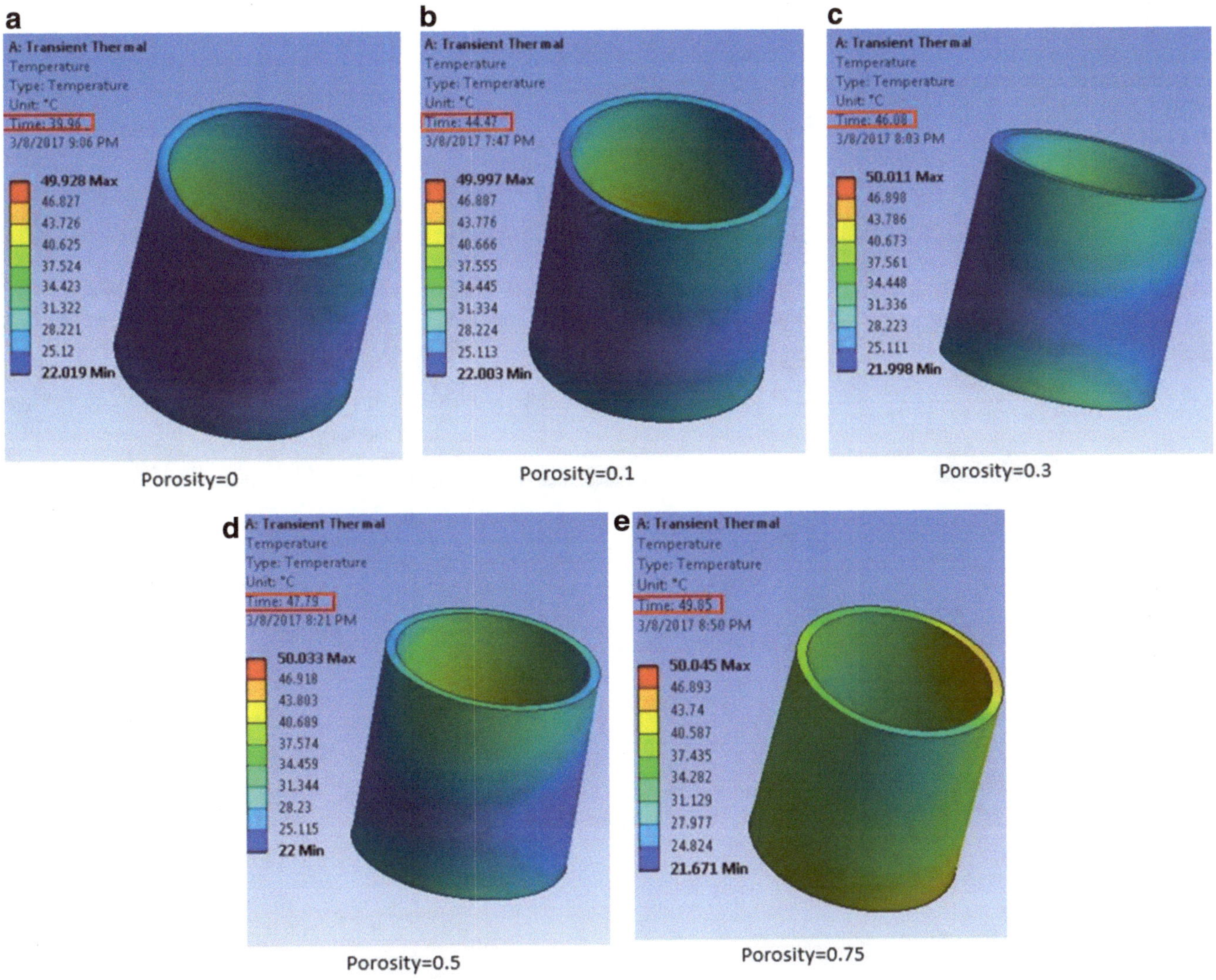

Fig. 8.6 Temperature distribution on plastic part at ejection, analysis with experimentally derived properties (**a**) porosity 0, (**b**) porosity 0.1, (**c**) porosity 0.3, (**d**) porosity 0.5, (**e**) porosity 0.75

Fig. 8.7 Comparative cooling time with porosity variation

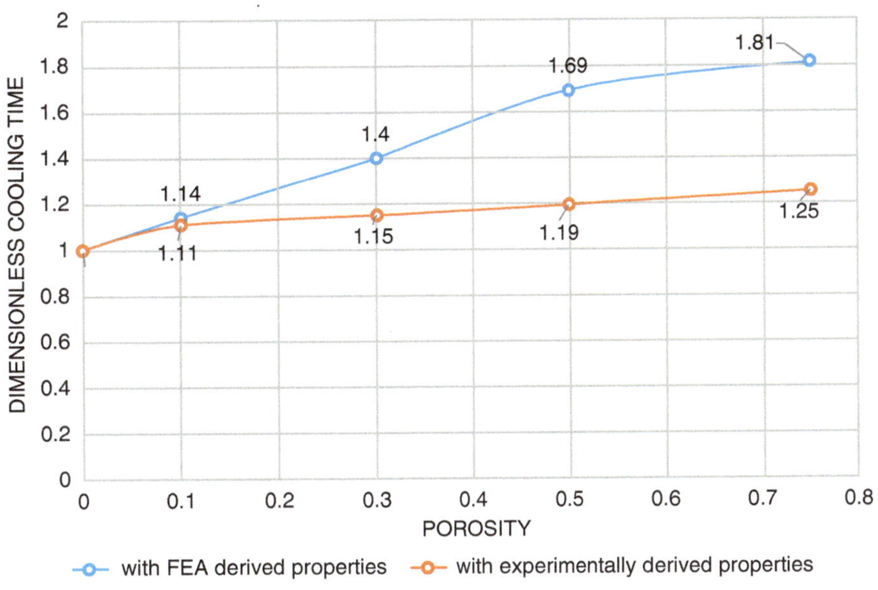

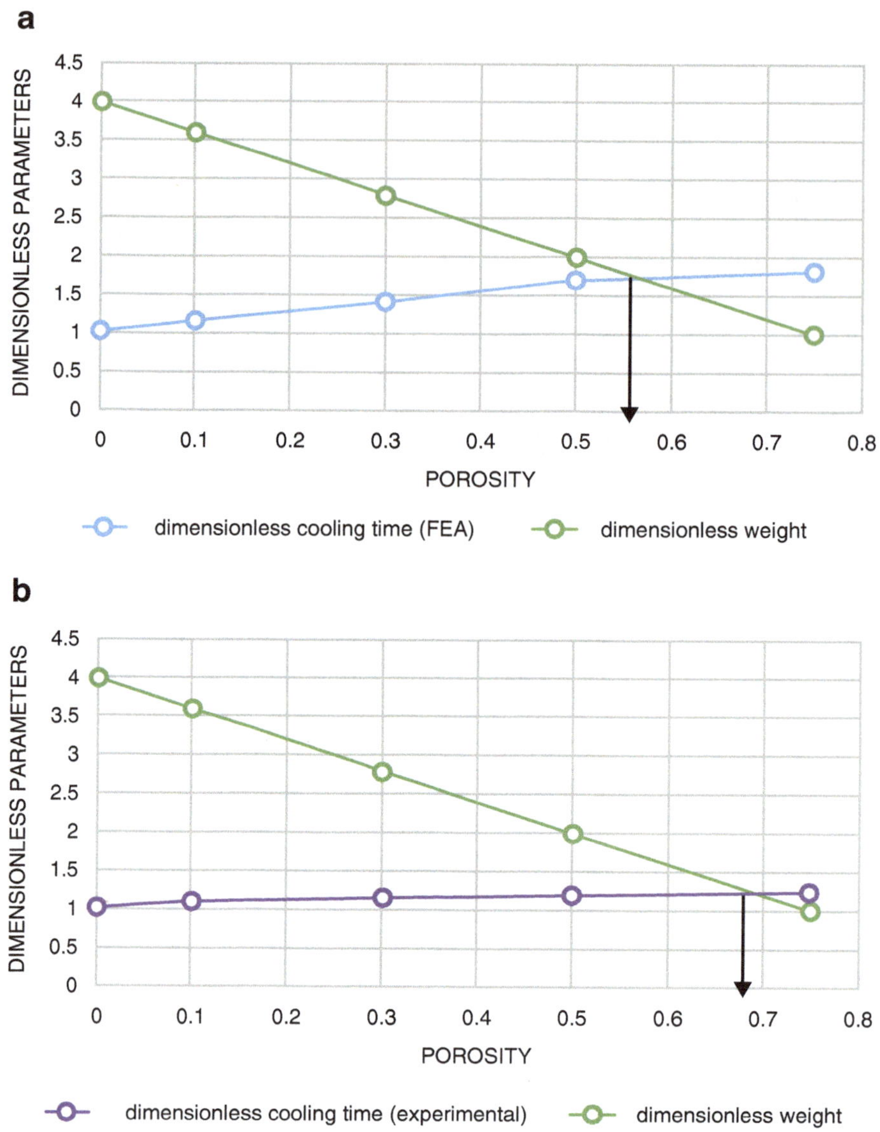

Fig. 8.8 Optimal porosity with cooling time and weight factors (**a**) FEA derived properties, (**b**) experimentally derived properties

of using 3D printed molds in injection molding industry. Reducing the weight of the mold by incorporating porosity will reduce the cost of production. Though there are other factors involved in the cost analysis, we are simplifying is by only the weight as an indication of cost. As discussed earlier, increasing porosity means increased cooling time. Hence, plotting the cooling time and weight in a graph, it is possible to identify a specific porosity which can provide the optimum benefit of minimum cooling time and minimum cost.

From Fig. 8.8a, the optimum design is obtained at porosity = 0.55 and from Fig. 8.8b, the optimum design is obtained at porosity = 0.68. From this study, we can conclude that mold material porosity in the range of 0.55–0.68 can be a good design solution for 3D printed molds with conformal cooling channels.

8.4 Conclusion and Future Work

In this study the effect of porosity on the thermal performance of 3D printed injection molds has been analyzed. With the increase of porosity, cooling time increases hence the thermal performance decreases. But high porosity is a positive factor in the cost perspective of 3D printing. Hence an optimization is required to find the most suitable design of injection molds for the benefit of injection molding industry business. The analysis in this study has been carried out based on 3D printed stainless steel material properties, obtained numerically and experimentally. There is about 5% variation between the material property values that leads to the variation in cooling time of injection molds with same porosities. However, considering the weight of mold as an indicator of cost and cooling time as the performance factor, a range of 0.55–0.68 porosity is identified as the optimal range of porosity for injection molds. The authors are currently working on identifying more specific results in terms of thermo-structural performance and detailed cost analysis.

References

1. Rosato, D.V., Rosato, M.G.: Injection Molding Handbook. Springer Science & Business Media, New York (2012)
2. Zheng, R., Tanner, R.I., Fan, X.-J.: Injection Molding: Integration of Theory and Modeling Methods. Springer Science & Business Media, Berlin (2011)
3. Dimla, D., Camilotto, M., Miani, F.: Design and optimisation of conformal cooling channels in injection moulding tools. J. Mater. Process. Technol. **164**, 1294–1300 (2005)
4. Sachs, E., et al.: Progress on tooling by 3D printing; conformal cooling, dimensional control, surface finish and hardness. In: Proceedings of the Eighth Annual Solid Freeform Fabrication Symposium, Austin (1997)
5. Sachs, E., et al.: Production of injection molding tooling with conformal cooling channels using the three dimensional printing process. Polym. Eng. Sci. **40**(5), 1232–1247 (2000)
6. Xu, R.X., Sachs, E.: Rapid thermal cycling with low thermal inertia tools. Polym. Eng. Sci. **49**(2), 305–316 (2009)
7. Xu, X., Sachs, E., Allen, S.: The design of conformal cooling channels in injection molding tooling. Polym. Eng. Sci. **41**(7), 1265–1279 (2001)
8. Xu, X., et al.: Designing conformal cooling channels for tooling. In: Solid Freeform Fabrication Proceedings (1998)
9. Ferreira, J., Mateus, A.: Studies of rapid soft tooling with conformal cooling channels for plastic injection moulding. J. Mater. Process. Technol. **142**(2), 508–516 (2003)
10. Meckley, J., Edwards, R.: A study on the design and effectiveness of conformal cooling channels in rapid tooling inserts. Technol. Interface J. **10**(1), 1–28 (2009)
11. Hopkinson, N., Dickens, P.: Conformal cooling and heating channels using laser sintered tools. In: Solid Freeform Fabrication Conference, Texas (2000)
12. Altaf, K., Rani, A. M. A., Raghavan, V.R.: Fabrication of circular and profiled conformal cooling channels in aluminum filled epoxy injection mould tools. In: National Postgraduate Conference (NPC), IEEE (2011)
13. Saifullah, A., Masood, S.: Finite Element Thermal Analysis of Conformal Cooling Channels in Injection Moulding. Engineers Australia, Brisbane (2007)
14. Saifullah, A., Masood, S., Sbarski, I.: New cooling channel design for injection moulding. In: Proceedings of the World Congress on Engineering (2009)
15. Saifullah, A., Masood, S., Sbarski, I.: Thermal–structural analysis of bi-metallic conformal cooling for injection moulds. Int. J. Adv. Manuf. Technol. **62**(1–4), 123–133 (2012)
16. Saifullah, A., Masood, S.H.: Optimum cooling channels design and thermal analysis of an Injection moulded plastic part mould. In: Materials Science Forum. Trans Tech Publications, Aedermannsdorf (2007)
17. Au, K., Yu, K.: A scaffolding architecture for conformal cooling design in rapid plastic injection moulding. Int. J. Adv. Manuf. Technol. **34** (5–6), 496–515 (2007)
18. Au, K., Yu, K.: Variable distance adjustment for conformal cooling channel design in rapid tool. J. Manuf. Sci. Eng. **136**(4), 044501 (2014)
19. Au, K., Yu, K., Chiu, W.: Visibility-based conformal cooling channel generation for rapid tooling. Comput. Aided Des. **43**(4), 356–373 (2011)
20. ó Gloinn, T., et al.: FEA simulation of conformal cooling within injection moulds. Int. J. Manuf. Res. **2**(2), 162–170 (2007)
21. Park, H.S., Pham, N.H.: Automatically generating conformal cooling channel design for plastic injection molding. Ann. DAAAM Proc. 539–541 (2007)
22. Park, H.-S., Pham, N.H.: Design of conformal cooling channels for an automotive part. Int. J. Automot. Technol. **10**(1), 87–93 (2009)
23. Wang, Y., et al.: Automatic design of conformal cooling circuits for rapid tooling. Comput. Aided Des. **43**(8), 1001–1010 (2011)
24. Jahan, S.A.: Optimization of conformal cooling channels in 3D printed plastic injection molds (2016)
25. Jahan, S.A., El-Mounayri, H.: Optimal Conformal Cooling Channels in 3D Printed Dies for Plastic Injection Molding. Procedia Manuf. **5**, 888–900 (2016, in press)
26. Jahan, S. A., et al.: Implementation of conformal cooling & topology optimization in 3D printed stainless steel porous structure injection molds. Procedia Manuf. **5**, 901–915 (2016, in press)
27. Wu, T., et al.: A framework for optimizing the design of injection molds with conformal cooling for additive manufacturing. Procedia Manuf. **1**, 404–415 (2015)
28. Isaacs, A., et al.: Thermal Modeling of a LightWeight Porous Structure (in press)
29. GmbH, E.: http://gpiprototype.com/images/PDF/EOS-StainlessSteel-PH1.pdf (2015). Retrieved on 15 Feb 2017

Chapter 9
ODS Coating Development Using DED Additive Manufacturing for High Temperature Turbine Components

Bruce S. Kang, Jaeyoon Kim, Eric Chia, Yang Li, and Minking Chyu

Abstract New generation of turbine blade coating using additive manufacturing (AM) technique to coat a layer of oxide dispersion strengthening (ODS) alloy on superalloy substrate is presented. A novel combined mechanochemical bonding (MCB) plus ball milling process is utilized to produce near spherical and uniform alloyed ODS powders. AM-assisted ODS coating by direct energy deposition (DED) method on MAR-247 substrate, with laser powers of 100 W, 150 W and 200 W were carried out. The ODS coated samples were then subjected to cyclic thermal loadings for over 1280 cycles. Corresponding Young's modulus measurements of ODS coating at various thermal loading cycles were conducted using a unique non-destructive micro-indentation testing method. Correlation of the measured Young's modulus with evolution of the ODS microstructures are studied. In particular, the presence of secondary gamma prime phase in the ODS coating after thermal cycles is noted. Test results revealed a thin steady durable alpha alumina oxide layer on the 200 W ODS sample. After 1280 thermal cycles, strong bonding at ODS/substrate interface is maintained for the 200 W ODS coated sample. Test results also showed stable substrate microstructures due to the protective ODS coating.

Keywords ODS coating • Additive manufacturing • Indentation • Oxide • Superalloy

9.1 Introduction

Nickel-based ODS alloys produced by mechanical alloying (MA) process attract great attention as advanced high temperature materials. This is because ODS alloys retain useful strength up to a relatively high temperature (1200 °C) due to direct strengthening of fine, uniformly dispersed and stable yttrium oxide particles which act as dislocation motion barrier and their ability to form protective Cr-oxide and Al-oxide surface layers during high temperature exposure [1]. In addition, the ability to exhibit improved scale adherence, decreased oxide growth rates, enhanced selective oxidation and decreased oxide grain size compared to corresponding non-ODS alloys made ODS alloys potentially suitable structural materials for components under high temperature and hostile service conditions. General processing of ODS alloys involve mechanical alloying, a powerful solid-state powder processing method, in which elemental alloy powders or oxide compounds are subjected to high energy ball milling or rod milling for long hours to produce homogeneously mixed, heavily deformed, cold welded composite particles. Typically after mechanical alloying, the mixed and alloyed powders are canned, degassed and consolidated by a number of compacting methods which include cold isostatic pressing (CIP), hot isostatic pressing (HIP) or hot extrusion followed by a final annealing and aging process at a selected high temperature procedure to develop stable and recrystallized microstructure with coarse and highly elongated grain structures and strengthening precipitates. However, over the years mechanical alloying has proven to be a very time and energy intensive process leading to high costs of manufacturing and powder contamination due to the complicated processes involved. To minimize the mentioned issues of mechanical alloying and to get better powder mixing and alloying, we adopted Hosokawa mechano-chemical bonding (MCB) technique that can disperse and mix elemental particles uniformly using only mechanical energy in a dry phase without needing solvents or external heating. This can avoid potential contaminations and reactions to the powders. It is an effective approach to blend alloying powders, forming composite particles consisting of hosting particles as core and small particles or fibers that are coated around the core particles [2]. During MCB processing, the starting powder mixtures are subjected to high compression, shear, and impact forces as they pass through a narrow gap in a high speed rotating device [2]

B.S. Kang (✉) • J. Kim • E. Chia
Department of Mechanical and Aerospace Engineering, West Virginia University, Morgantown, WV 26506, USA
e-mail: Bruce.Kang@mail.wvu.edu

Y. Li • M. Chyu
Department of Mechanical Engineering and Materials Science, University of Pittsburgh, Pittsburgh, PA 15261, USA

© The Society for Experimental Mechanics, Inc. 2018

J. Wang et al. (eds.), *Mechanics of Additive and Advanced Manufacturing, Volume 9*,
Conference Proceedings of the Society for Experimental Mechanics Series, DOI 10.1007/978-3-319-62834-9_9

(typically around 4000 rpm), where the elemental particles are dispersed, mixed, shaped, and bonded together forming composite particles of the starting materials. The MCB technology has been utilized to make various composite particles used in technology areas of functional gradient materials (FGM), batteries, cermets, fuel cells, polymers, cosmetics, and pharmaceuticals [2].

Our earlier studies, utilizing MCB technique to blend nickel-based ODS alloy powders, showed the ability of homogenously dispersing yttrium oxide particles on the base hosting particles to form a composite structure generating a nano-sized Y-Al-O enriched film at each hosting particle [2]. A homogeneous distribution of yttrium oxide dispersion provides the material more stability at elevated temperature by which the nano-sized yttrium oxide particles are expected to form a dispersion throughout the core particles. This enhances formation of stable and adherent protective oxide scales with good mechanical properties which are very important aspects to withstand degradation from reaction with gases and condensed products at high temperature.

Recently, we have developed a new processing route of producing alloyed ODS composite powders based on a combined MCB and ball milling (BM) methodology [2, 3]. ODS alloy samples were produced using cold isostatic pressing (CIP) with excellent mechanical strength retention after prolonged cyclic thermal loading testing (up to 1100 °C) [4]. In this paper, we A key research effort in this paper is to develop ODS composite powders suitable for 3D AM of advanced turbine component It should be emphasized that current commercially available ODS powders are not suitable for fabrication of structural components using 3D AM technique mainly due to brittleness of the ODS particles with uneven nano-sized yttrium oxide dispersion.

9.2 ODS Powder Fabrication

In order to fabricate alloyed ODS composite powders that are near spherical shape with proper particle size range suitable for AM applications. The MCB-processed ODS elemental mixed particles were then subjected to high energy ball milling (BM). Various selections of Process Control Agent (PCA) and Ball to Powder Ration (BPR) were used. Based on these optimization runs, we have decided to use 15:1 BPR with a two-stage ball milling route (i) 20 h at 400 rpm, and (ii) then 20 h at 150 rpm. Figure 9.1 shows an XRD plots of serval process conditions. As shown, 15:1 BPR ball milled for 15 h is a suitable procedure to produce alloyed ODS particles, also noted is that XRD spectra for 30:1 BPR at 15 h is similar to 15:1 BPR at 15 h. Figure 9.2 shows SEM images for the 15:1 BPR alloyed ODS powders with mostly spherical particles in the desirable size range. As shown, the top left image shows an even distribution of the ODS particles and the images on the right also shows there are few particles less than 10 micrometers. The increased speed at 20 h demonstrates the effect of the sphericity of the particles and the lowered speed for another 20 h promoted cold welding, thus the increase of particle sizes. AM trial runs at Pitt using this production run of ODS particles was very successful.

Fig. 9.1 XRD plots of ODS particles, MCB only and MCB + BM (different BPR ratio)

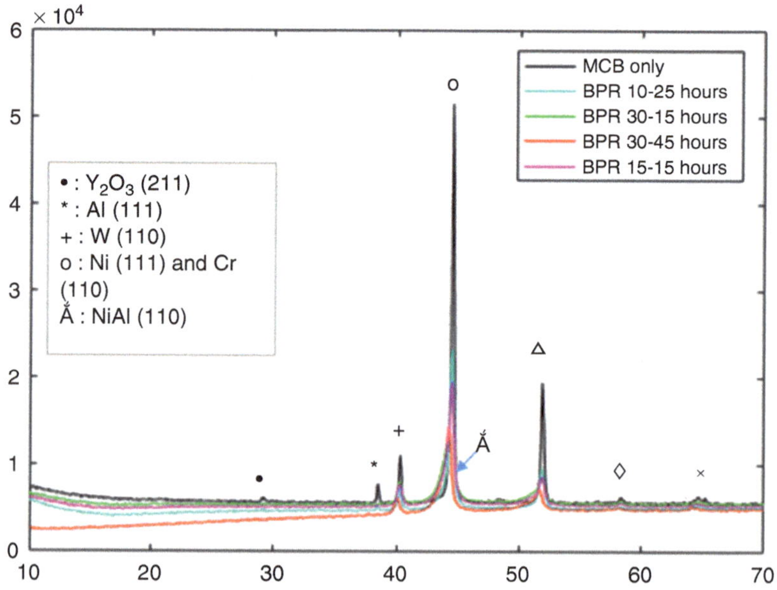

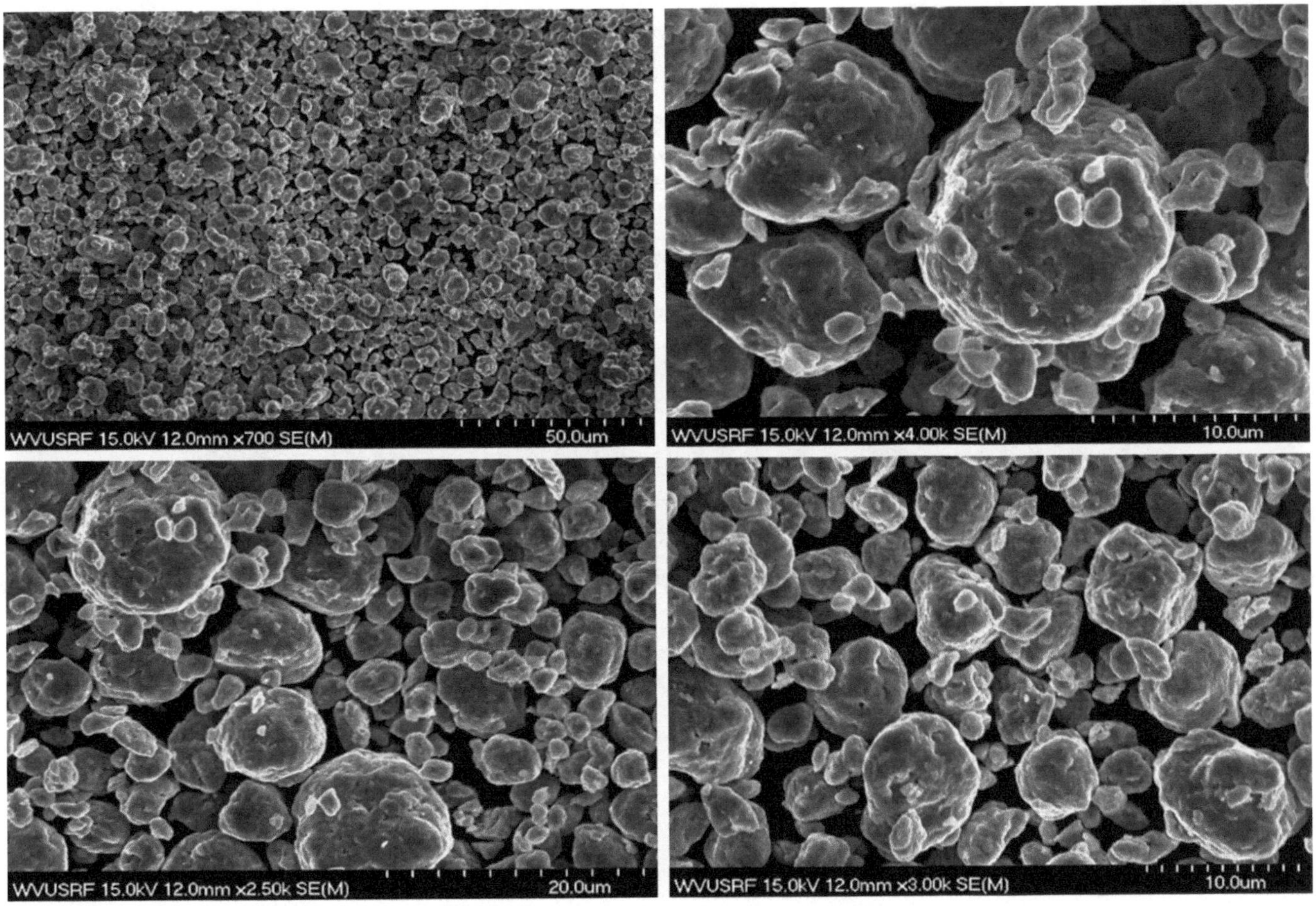

Fig. 9.2 SEM image of ODS particles ball milled with BPR of 15:1 initially for 20 h at 400 rpm and 20 h at 150 rpm

9.3 Thermal Cyclic Tests of ODS Coupons

ODS coating was done using the AM equipment at Pitt (LENS450, Optomec Inc.) and WVU ODS powders. The LENS450 has a built chamber of 100 × 100 × 100 mm (4.0″ × 4.0″ × 4.0″) which is purged by Argon gas to maintain low oxygen level (below 10 ppm) to ensure there is very minimal/no impurity during the deposition process. The component is build layer by layer through a well-developed software that controlled various processing parameters that ensure geometric and mechanical integrity. The metal powder is delivered using Optomec's proprietary powder feed system that has been optimized to dispense small quantities of metal powder precisely. The current LENS450 is equipped with two powder feeders, which allows the user to mix two different metal powders at varying ratio, eventually presents a great opportunity for metal powder alloy development studies. Figure 9.3 shows Am-assisted ODS coating on MM247 superalloy substrate, excellent interface adhesion to the substrate is noted. The as AM-processed ODS specimens were then subjected to thermal cyclic testing, which is a standard lab testing for gas turbine components. The thermal cyclic oxidation testing was conducted in a tube furnace at 1100 °C in air. The arrangement of the thermal cyclic testing was done in a manner that the sample is moved in to the furnace slowly for about 15 min and stayed in at 1100 °C for 45 min and slowly moved out for 15 min and stayed 45 min at room temperature, which made it to be one cycle and continued automatically for the next cycles. It is also important to note that, after thermal cyclic loadings (>1280 cycles), strong adhesion between the ODS coating and substrate remains. To the best of our knowledge, this is the first time in material research community to demonstrate AM-assisted ODS coating/substrate system can survive under high temperature condition (1100 °C).

The performance of oxide scales developed on the ODS specimens were investigated by an in-house developed nondestructive micro-indentation technique [5], which consists of a multiple-partial unloading procedure during indentation testing, resulting in a load-depth sensing indentation system capable of determining elastic modulus of metallic alloys with flat, tubular, or curved architectures [5]. Test results showed consistent and correct elastic modulus values when performing indentation tests on standard alloys such as steel, aluminum, bronze, and single crystal superalloys having a precision with

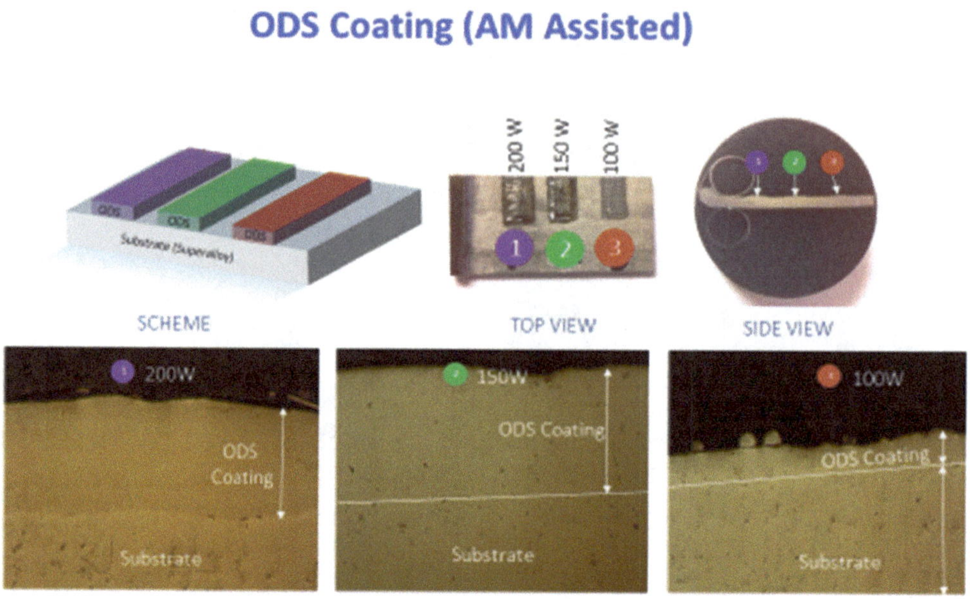

Fig. 9.3 AM-assisted ODS coating on MAR-247 superalloy substrate

Table 9.1 Measured surface mechanical property (Young's Modulus)

	200 W	150 W	100 W	Substrate
As-received	174.4	173.8	227.6	126.5
15 cycles	141.5	80.9		100.2
40 cycles	132.1	85.5		109.1
	143.0	86.8		
80 cycles	113.2	59.4	57.3	58.7
	113.0	58.4	57.4	61.0
160 cycles	110.3	68.8	62.4	
	135.4	65.4	61.2	
240 cycles	210.0	72.0		
	213.5	88.8		
360 cycles	197.3	123.4		123
	181.5	136.1		125.7
480 cycles	250.5	100.5	63.8	124.9
	229.4	98.2	59.0	141.7
600 cycles	183.0	84.7	163.6	124.6
	225.0	96.5	139.0	140.2
720 cycles	201.5	176.5	116.1	155.8
	205.4	200.4		168.9
1040 cycles	194.6			99.3
	188.7			92.9
1280 cycles	197.6	204.4	101	
	178.4	154.4	113	

coefficient of variation less than 4.5%. As shown, preliminary results showed that the AM-assisted ODS specimen has retained the mechanical strength after 1048 cycles. Table 9.1 shows the measured surface mechanical property (Young's modulus) at discrete thermal cycles. As shown, due to the instability of surface oxidation, performance of 100 W ODS coating is not desirable. For the case of 150 W ODS coating, surface oxide layer formation is also unstable, there's spallation and regrowth of oxide layer and the measured surface layer stiffness (i.e. Young's modulus) varied substantially location to location. However, the initial substantial increase and stability of surface Young's modulus of the 200 W ODS coating is noted. Figures 9.4, 9.5 and 9.6 show SEM/EDX analyses of the 200 W ODS coating. The results shows gradual mixed oxide formation of mainly Al_2O_3 and NiO oxides. The thermal cyclic testing is on-going until failure of the ODS coating is

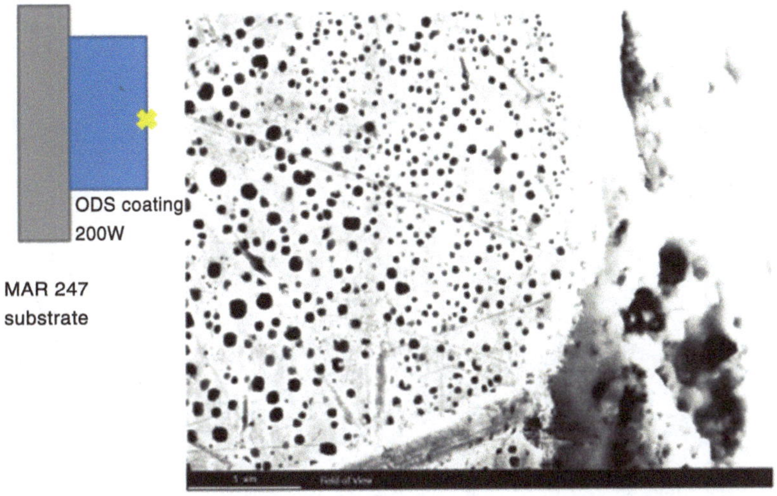

Fig. 9.4 SEM image of ODS microstructure at 1280 thermal cycles (next to surface oxide)

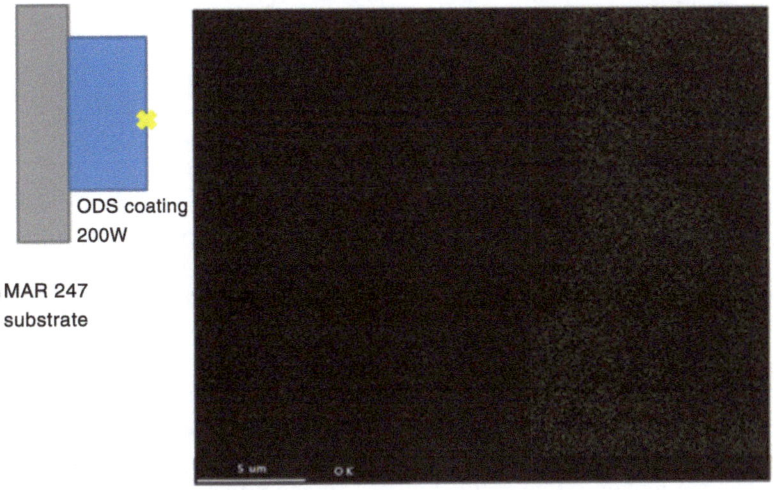

Fig. 9.5 EDX image of oxygen which indicates the presence of alpha alumina oxide layer on the ODS coating surface at 1280 thermal cycles

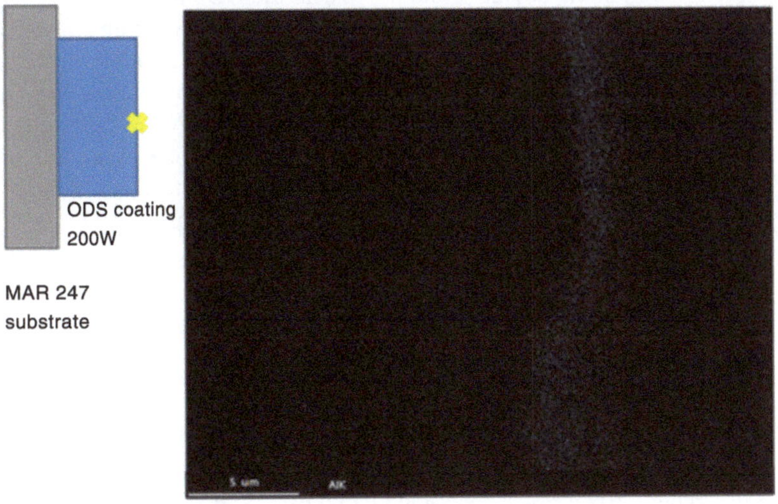

Fig. 9.6 EDX image of aluminum which indicates the presence of alpha alumina oxide layer on the ODS coating surface at 1280 thermal cycles

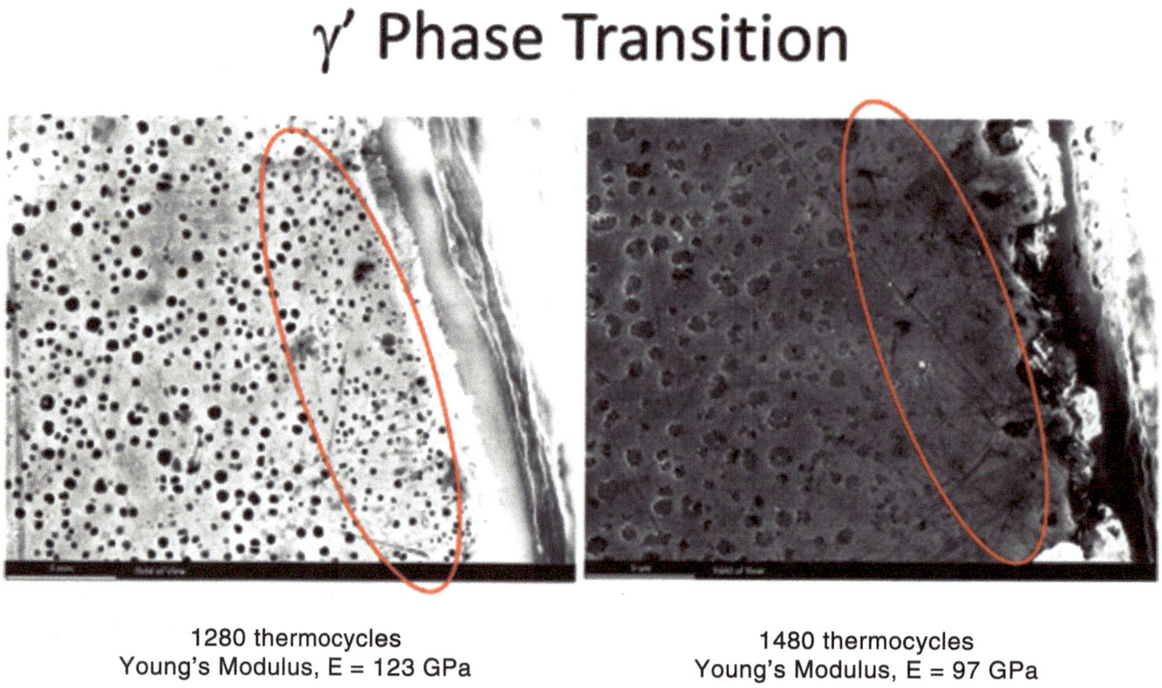

Fig. 9.7 150 W ODS coating specimen. SEM images of near surface coating microstructure at 1280 and 1480 thermal cycles. Depletion of gamma double prime phase at 1480 thermal cycles is noted

observed. As shown, at 1280 cycles, relative stable and high Young's modulus of the 200 W AM-assisted ODS coating are observed. This is due to the stable formation of top Al_2O_3 oxide layer and the gamma double prime phases at the ODS coating matrix. This is a strong evidence that indeed the AM-assisted ODS coating is a structural coating. Under long-term thermal loading, it has a durable surface Al_2O_3 oxide layer such that aluminum in the ODS matrix diffuses to the surface interacting with oxygen to form the Al_2O_3 oxide layer. Because of the uniform dispersion of Yttrium oxide which control the kinetic diffusion of aluminum, the formation of the surface oxide Al_2O_3 is slow and durable, i.e. thus far, at 1280 thermal cycles, we have not observed thick Al_2O_3 oxide and cracking of the Al_2O_3 oxide layer. It should be emphasized that without the kinetic diffusion of the aluminum from substrate (not possible), the gamma prime and gamma double prime phases in the substrate remain stable (very slow coarsening effect, as shown in Fig. 9.6) and thus mechanical strength of the substrate remain intake, i.e. the turbine blade will have long-term structural stability).

For comparison, as shown in Table 9.1, there's drop of measured Young's modulus of the 150 W AM ODS coating at 1280 thermal cycles. Upon closer examination, we observed that the decrease of Young's modulus can be related to the depletion of gamma double prime phase near the ODS coating surface, as shown in Fig. 9.7. After long-term thermal cyclic testing, oxide layer on the 150 W ODS coating specimen became unstable that allowed oxygen to penetrate down to the ODS coating at discrete locations. As a consequence gamma double prime phase in those locations started to show coarsening effect and will be disintegrated, as shown in the marked region of Fig. 9.7. This preliminary test results clearly indicated the need to optimize AM-assisted fabrication parameters, in particular, the laser power. Thus far, use of 200 W laser power for AM-assisted ODS coating seems to be the optimized one. Also, due to the protective ODS coating with no possibility of oxygen diffusion to the substrate alloy, the microstructures of the substrate is preserved subjected only to thermal cyclic effect with some small increase of the size of gamma prime phase, as shown in Fig. 9.8 at 1280 thermal cycles.

9.4 Conclusion

Utilizing additive manufacturing technique, high temperature resistance structural oxide dispersion strengthening (ODS) coating on superalloy substrate is presented. A novel combined mechanochemical bonding (MCB) plus ball milling process is utilized to produce near spherical ODS powders. AM-assisted ODS coating by direct energy deposition (DED) method on MAR-247 substrate, with laser powers of 100 W, 150 W and 200 W were carried out. Test results revealed a thin steady

Fig. 9.8 SEM image of MAR 247 substrate at 1280 thermal cycles

durable alpha alumina oxide layer on the 200 W ODS sample. After 1280 thermal cycles, strong bonding at ODS/substrate interface is maintained and stable substrate microstructures are preserved due to the protective ODS coating.

Acknowledgement This research is supported by US Department of Energy, National Energy Technology Laboratory (NETL), University Turbine System Research Program, Award Number: DE-FE0025793, 10/1/2015 to 9/30/2018. The encouragement and support by Richard Dennis and Seth Lawson, Program Managers, are very much appreciated.

References

1. Weinbruch, S., Anastassiadis, A., Ortner, H.M., Martinz, H.P., Wilhartitz, P.: On the mechanism of high temperature oxidation of ODS superalloys: significance of yttrium depletion within the oxide scales. Oxid. Met. **51**, 111–128 (1999)
2. Suryanarayana, C.: Mechanical Alloying and Milling. Marcel Dekker, New York (2004)
3. Ma, L., Kang, B.S.-J., Alvin, M.A., Huang, C.C.: Characterization of oxide-dispersion-strengthened (ODS) alloy powders processed by mechano-chemical-bonding (MCB) and balling milling (BM). Kona. Powder. Part. J. **31**, 146–155 (2014)
4. Amare, B.N.: Mechanical and microstructure study of nickel-based ods alloys processed by mechano-chemical bonding and balling milling. Ph. D. dissertation, Mechanical and Aerospace Engineering Department, West Virginia University (2015)
5. Feng, C., Tannenbaum, J.M., Kang, B.S., Alvin, M.A.: A load-based multiple-partial unloading micro-indentation technique for mechanical property evaluation. Exp. Mech. **50**(6), 737–743 (2010)

Chapter 10
Processing and Characterization of Ti64/AZ31 Multilayered Structure by Roll Bonding

Chin Shih Hsu and Qizhen Li

Abstract Ti64 titanium alloy sheets and AZ31 magnesium alloy sheets were stacked in an alternating order and rolled to form three types of roll-bonded multilayered materials. The microstructure of these roll-bonded samples was observed using optical microscope, and the Vickers hardness data were also obtained. The thickness of the AZ31 layers was severely reduced, while the thickness of Ti64 layers was only slightly changed. Compared with the as-received Ti64 titanium alloy and AZ31 magnesium alloy, the Vickers hardness of Ti64 layers only slightly deviated from that of the as-received state, while AZ31 layers was refined equiaxed grains had higher hardness than that of the as-received state.

Keywords Ti64 titanium alloy • AZ31 magnesium alloy • Microstructure • Microhardness • Rolling

10.1 Introduction

With the pursuit of better properties, researchers are exploring different types of materials. One attractive type of materials is the combination of multiple different materials such as pure metal [1–3], alloy [4, 5], two different metals [6–8], three different metals [9–11], and composites [12–16] through roll bonding [17]. To date, there is still a lack of research results about multilayered materials with alternating titanium layer and magnesium layer that are processed using roll bonding technique. It is meaningful to investigate the production and properties of roll-bonded titanium/magnesium multilayered materials since both are light and the materials would be attractive as lightweight and high strength materials. This work produced roll-bonded multilayered materials with alternating Ti64 titanium alloy layer and AZ31 magnesium alloy layer and performed microstructure and mechanical characterizations of these multilayered materials.

10.2 Experimental Methods

This study utilized two types of metals, i.e. Ti6Al4V titanium alloy (Ti64) and Mg3Al1Zn magnesium alloy (AZ31), to produce different multilayered materials through the following steps. At the first step, Ti64 was heated at 500 °C and rolled to the sheets with a thickness of ~0.4 mm, and AZ31 was heated at 500 °C and rolled to the sheets with a thickness of ~0.5 mm. At the second step, the Ti64 and AZ31 sheets were cut into pieces with a width of 20 mm and a length of 40 mm, ground with sandpapers to remove oxide layers, and degreased and cleaned using acetone. As the third step, the cleaned sheets were stacked and rolled to produce three kinds of multilayered materials with the stacking sequences of Ti64/AZ31/Ti64, Ti64/AZ31/Ti64/AZ31/Ti64, and Ti64/AZ31/Ti64/AZ31/Ti64/AZ31/Ti64 respectively. For the case of Ti64/AZ31/Ti64, the stack was heated at 500 °C for 1 min and then rolled for one pass with an actual rolling reduction of 25%. For the case of Ti64/AZ31/Ti64/AZ31/Ti64, the stack was heated at 500 °C for 1 min and then rolled for one pass with an actual rolling reduction of 25%. For the case of Ti64/AZ31/Ti64/AZ31/Ti64/AZ31/Ti64, two pieces of rolled Ti64/AZ31/Ti64 were used to sandwich a 0.5 mm thick AZ31 layer to be heated at 500 °C for 1 min and then rolled for one pass with an actual total rolling reduction of 38%. Each type of rolled sample was cut along a plane perpendicular to the layer interfaces and the rolling direction (RD) to observe the transverse-direction (TD) cross-section, and also was cut along a plane perpendicular to the layer interfaces and parallel to the RD to observe the RD cross-section. The cut samples were mounted through cold

C.S. Hsu • Q.Z. Li (✉)
School of Mechanical and Materials Engineering, Washington State University, Pullman, WA 99164, USA
e-mail: qizhen.li@wsu.edu

© The Society for Experimental Mechanics, Inc. 2018

J. Wang et al. (eds.), *Mechanics of Additive and Advanced Manufacturing, Volume 9*,
Conference Proceedings of the Society for Experimental Mechanics Series, DOI 10.1007/978-3-319-62834-9_10

mounting, ground using sandpapers, polished using alumina suspension, and etched for microstructure observations under optical microscope (OM). The as-received Ti64 and the as-received AZ31 were also observed for comparison. Vickers hardness data were obtained for both Ti64 layer and AZ31 layer for each sample through microhardness testing.

10.3 Results and Discussion

Figure 10.1 shows the internal microstructure of the as-received AZ31 and the as-received Ti64. The grains for the as-received materials are equiaxed. Figure 10.2 shows the optical micrographs of the RD and TD cross-sections of each type of roll-bonded multilayered samples. For all the roll-bonded samples, the thickness of the AZ31 layers was severely reduced, while the thickness of Ti64 layers was only slightly changed. The possible reason for this phenomenon is that Ti64 has much higher strength than AZ31. Among the three types of produced materials, the Ti64/AZ31/Ti64/AZ31/Ti64 sample showed the smallest thickness reduction of AZ31 layers, and the Ti64/AZ31/Ti64/AZ31/Ti64/AZ31/Ti64 sample showed the largest thickness reduction of AZ31 layers. All the roll-bonded samples have almost flat interfaces between AZ31 layer and Ti64 layer. The waviness of the AZ31/Ti64 interfaces in the Ti64/AZ31/Ti64/AZ31/Ti64 sample is the highest among these three types of miltilayered materials.

Figures 10.3 and 10.4 report the microstructure of Ti64 layer and AZ31 layer along both RD and TD cross-sections for the roll-bonded Ti64/AZ31/Ti64, Ti64/AZ31/Ti64/AZ31/Ti64, and Ti64/AZ31/Ti64/AZ31/Ti64/AZ31/Ti64 samples.

Figure 10.3 shows that the grains in the Ti64 layers were elongated along the rolling direction and slightly elongated along the transverse direction. For the Ti64/AZ31/Ti64/AZ31/Ti64 and Ti64/AZ31/Ti64/AZ31/Ti64/AZ31/Ti64 samples, the microstructure of the Ti64 layers is similar to that of the as-received Ti64.

The grains in the AZ layers were equiaxed and finer than the as-received AZ31. Among the three types of roll-bonded samples, the Ti64/AZ31/Ti64/AZ31/Ti64/AZ31/Ti64 sample with the largest thickness reduction of AZ31 layers has the finest AZ31 grains, while the Ti64/AZ31/Ti64/AZ31/Ti64 sample with the smallest thickness reduction of AZ31 layers has the coarsest AZ31 grains. This phenomenon may be due to the relation between the melting temperatures of the involved materials and the heating temperature (500 °C) before rolling. The ratio between the heating temperature and the melting temperature of AZ31 is about 0.84, while this ratio for Ti64 is about 0.40. Even though the heating duration is short (i.e., 1 min), there is complete recrystallization and grain growth does not happen yet in the AZ layers and the microstructure shows fine equiaxed grains. Since the heating temperature just reaches the recrystallization temperature for Ti64 and the heating duration is only 1 min, the Ti64 layers just started the recrystallization process and kept the deformed structure.

Table 10.1 lists the Vickers hardness of the as-received Ti64, the as-received AZ31, the Ti64 layer and the AZ31 layer of the roll-bonded samples (i.e. Ti64/AZ31/Ti64, Ti64/AZ31/Ti64/AZ31/Ti64, and Ti64/AZ31/Ti64/AZ31/Ti64/AZ31/Ti64) along both RD and TD cross-sections. The average Vickers hardness for the as-received Ti64 is 164HV and that for the as-received AZ31 is 41HV, and these data indicate that the as-received Ti64 is much stronger than the as-received AZ31 and it is reasonable that the deformation was mainly concentrated in the AZ31 layers during rolling. The average Vickers hardness in the Ti64 layers is in the range of 160–178 HV. The Vickers hardness data of the Ti64 layers in the three types of

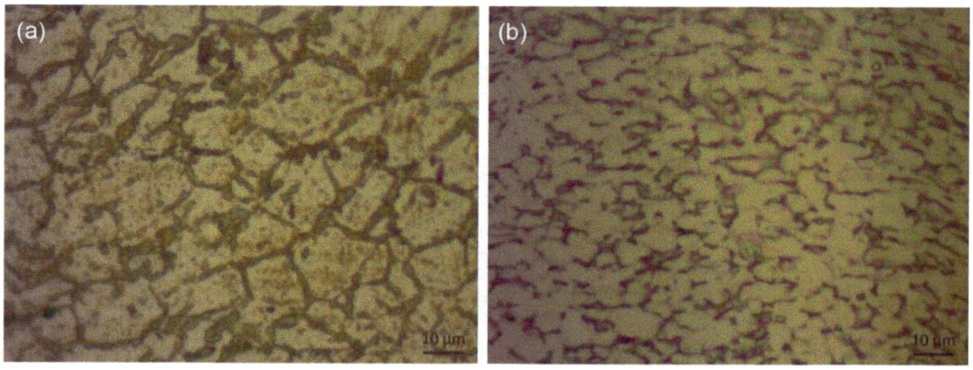

Fig. 10.1 Optical micrographs of the microstructure of (**a**) as-received AZ31 magnesium alloy and (**b**) as-received Ti64 titanium alloy

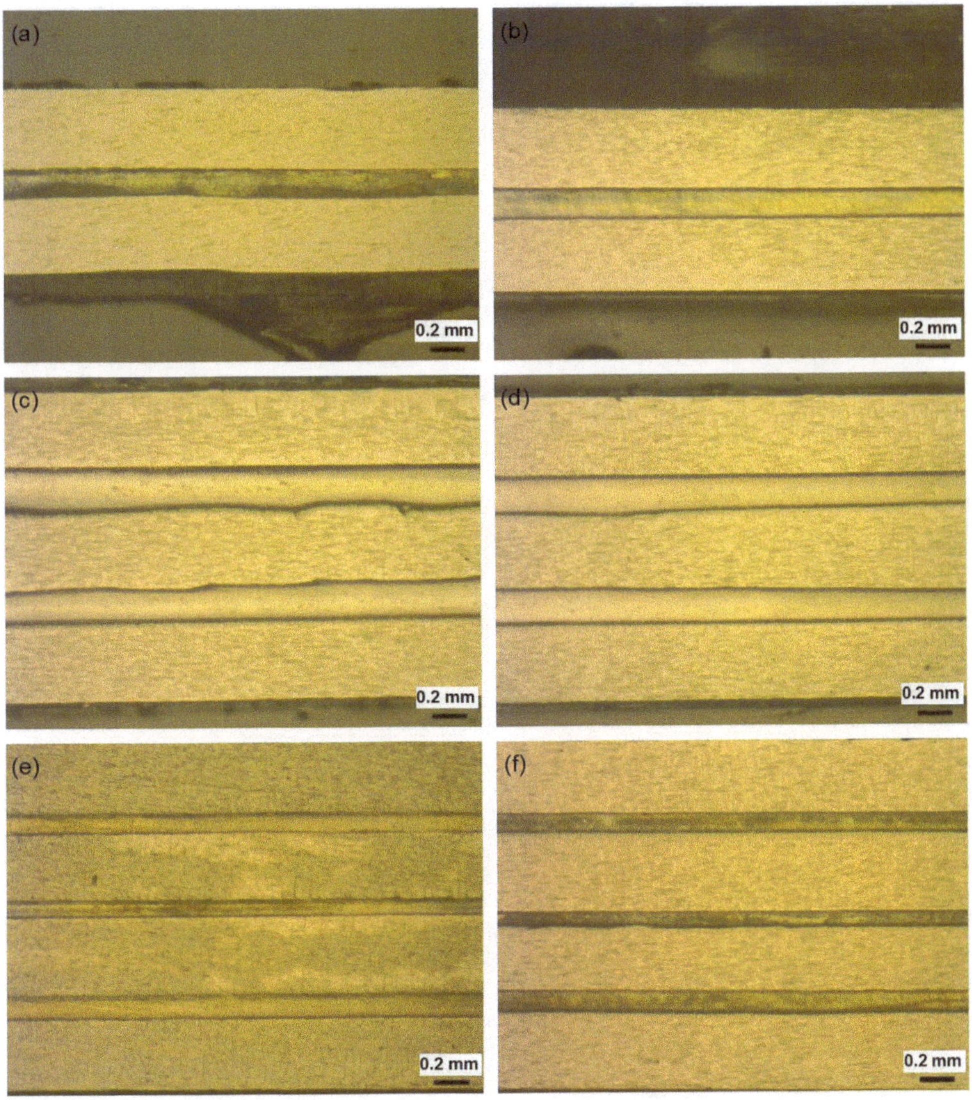

Fig. 10.2 Optical micrographs of the roll-bonded multilayered samples. (**a**) RD and (**b**) TD cross-sections of the Ti64/AZ31/Ti64 sample; (**c**) RD and (**d**) TD of the Ti64/AZ31/Ti64/AZ31/Ti64 sample; (**e**) RD and (**f**) TD of the Ti64/AZ31/Ti64/AZ31/Ti64/AZ31/Ti64 sample

roll-bonded samples are not significantly different from each other, and these data are also not significantly different from the data of the as-received Ti64. Due to the recrystallization in the AZ layers, the average grain sizes of the AZ layers for the roll-bonded samples are drastically decreased from the size of the as-received AZ31. The average Vickers hardness of the AZ layers is in the range of 36–70 HV. The average Vickers hardness of the AZ31 layers in the Ti64/AZ31/Ti64/AZ31/Ti64/AZ31/Ti64 sample is the highest among the three types of roll-bonded materials.

10.4 Conclusion

Three types of roll-bonded multilayered materials (i.e. Ti64/AZ31/Ti64, Ti64/AZ31/Ti64/AZ31/Ti64, and Ti64/AZ31/Ti64/AZ31/Ti64/AZ31/Ti64) were produced through stacking Ti64 sheets and AZ31 sheets in three types of sequences and then heated at 500 °C for 1 min and rolled with the total actual reduction percentages of 25%, 25%, and 38% respectively. These samples were characterized microstructurally and mechanically using optical microscope and microhardness tester. The results show that (a) the deformation was primarily concentrated in the AZ31 layers during rolling, (b) there is very little deformation

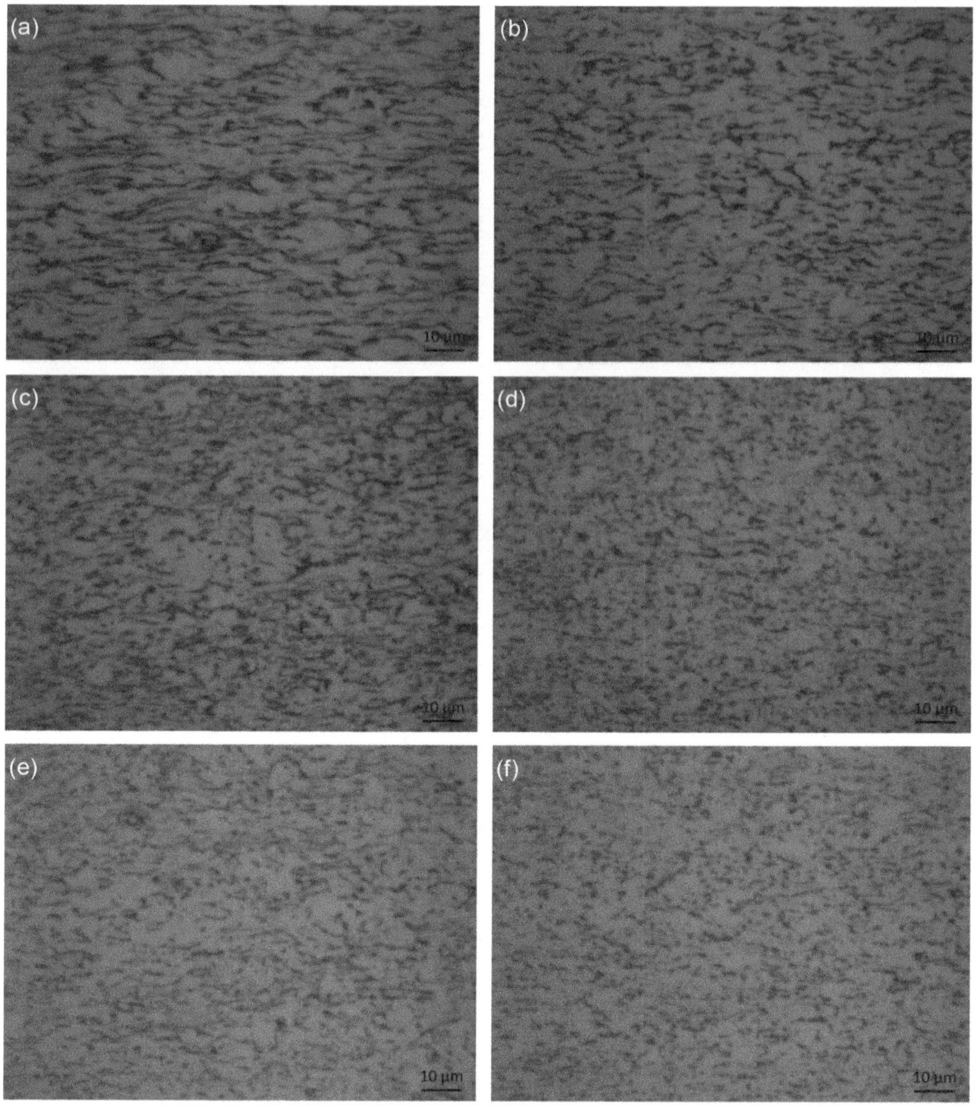

Fig. 10.3 Optical micrographs of the Ti64 layer of the roll-bonded multilayered samples. (**a**) RD and (**b**) TD cross-sections of the Ti64/AZ31/Ti64 sample; (**c**) RD and (**d**) TD of the Ti64/AZ31/Ti64/AZ31/Ti64 sample; (**e**) RD and (**f**) TD of the Ti64/AZ31/Ti64/AZ31/Ti64/AZ31/Ti64 sample

in the Ti64 layer since Ti64 is much stronger than AZ31, (c) the AZ layer has refined equiaxed grains since the heating temperature is ~84% of magnesium's melting temperature and complete recrystallization happened without grain growth, (d) the Vickers hardness in the Ti64 layer is not significantly different from each other for the three types of roll-bonded samples, and it is also not significantly different from that of the as-received Ti64, (e) the Vickers hardness of the AZ31 layer is increased compared with that of the as-received AZ31, and (f) the roll-bonded sample with the largest thickness reduction of the AZ31 layers has the highest Vickers hardness data in its AZ31 layers among the three types of materials.

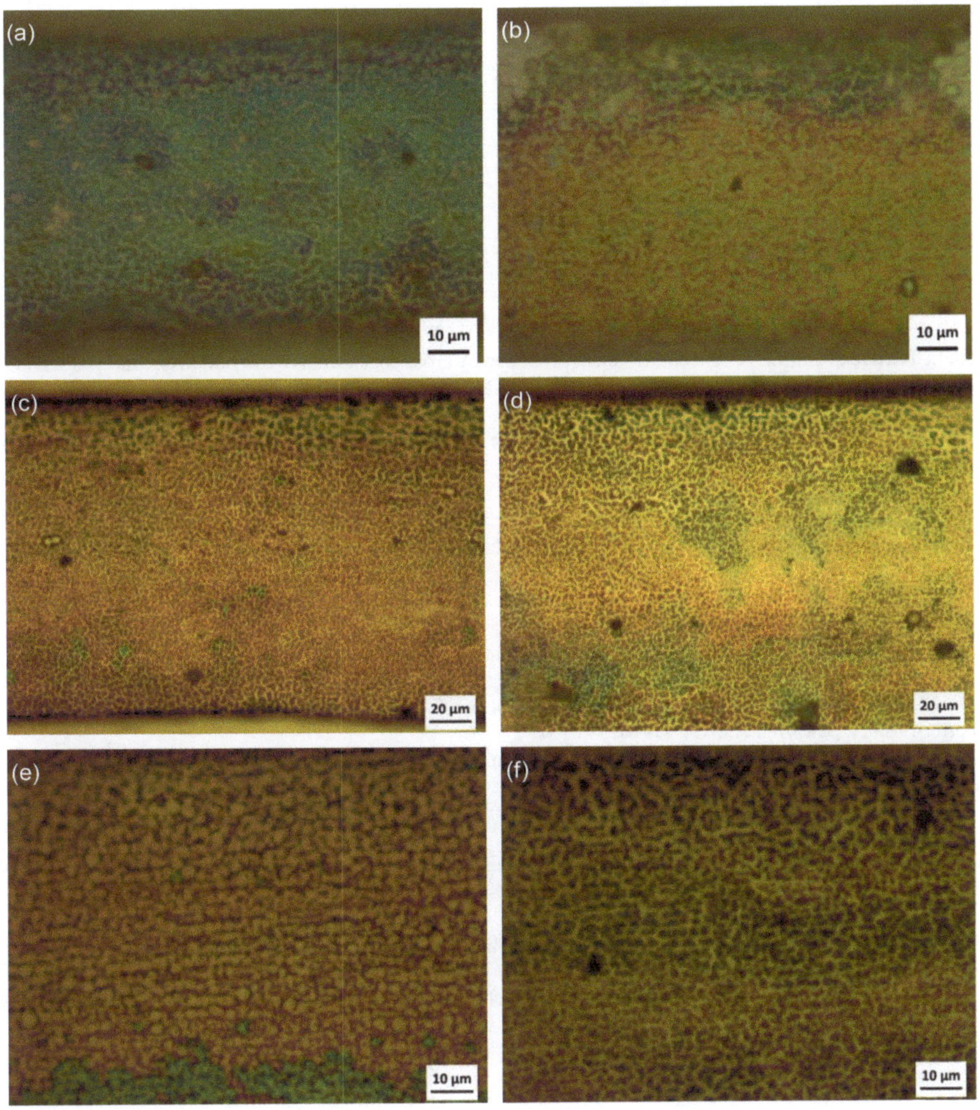

Fig. 10.4 Optical micrographs of the AZ31 layer of the roll-bonded multilayered samples. (**a**) RD and (**b**) TD cross-sections of the Ti64/AZ31/Ti64 sample; (**c**) RD and (**d**) TD of the Ti64/AZ31/Ti64/AZ31/Ti64 sample; (**e**) RD and (**f**) TD of the Ti64/AZ31/Ti64/AZ31/Ti64/AZ31/Ti64 sample

Table 10.1 Vickers hardness of the as-received Ti64, the as-received AZ31, the roll-bonded Ti64/AZ31/Ti64 sample, the roll-bonded Ti64/AZ31/Ti64/AZ31/Ti64 sample, and the roll-bonded Ti64/AZ31/Ti64/AZ31/Ti64/AZ31/Ti64 sample

Materials	Vickers hardness (HV)
As-received T64	164 ± 7
As-received AZ31	41 ± 3
Ti64/AZ31/Ti64, RD	164 ± 5(Ti64 layer), 46 ± 4(AZ31 layer)
Ti64/AZ31/Ti64, TD	173 ± 14(Ti64 layer), 49 ± 3(AZ31 layer)
Ti64/AZ31/Ti64/AZ31/Ti64, RD	175 ± 6(Ti64 layer), 42 ± 4(AZ31 layer)
Ti64/AZ31/Ti64/AZ31/Ti64, TD	171 ± 10(Ti64 layer), 36 ± 2(AZ31 layer)
Ti64/AZ31/Ti64/AZ31/Ti64/AZ31/Ti64, RD	178 ± 7(Ti64 layer), 63 ± 15(AZ31 layer)
Ti64/AZ31/Ti64/AZ31/Ti64/AZ31/Ti64, TD	160 ± 7(Ti64 layer), 70 ± 13(AZ31 layer)

Acknowledgement The support for the research from the Army Research office and Washington State University is greatly appreciated.

References

1. Bonnot, E., et al.: Microstructure and texture evolution during the ultra grain refinement of the Armco iron deformed by accumulative roll bonding (ARB). Mater. Sci. Eng. A. **561**, 60–66 (2013)
2. Miyajima, Y., et al.: Dislocation density of pure copper processed by accumulative roll bonding and equal-channel angular pressing. Mater. Charact. **104**, 101–106 (2015)
3. Li, Q.Z., Tian, B.: Mechanical properties and microstructure of pure polycrystalline magnesium rolled by different routes. Mater. Lett. **67**(1), 81–83 (2012)
4. Azzeddine, H., et al.: Texture evolution of Fe–Ni alloy sheet produced by cross accumulative roll bonding. Mater. Charact. **97**, 140–149 (2014)
5. Schwarz, F., Eilers, C., Krüger, L.: Mechanical properties of an AM20 magnesium alloy processed by accumulative roll-bonding. Mater. Charact. **105**, 144–153 (2015)
6. Hsu, C.S., Zou, N., Li, Q.Z.: Microstructure and mechanical properties of ti/az31 multi-layered materials processed by accumulative roll bonding. In: Shaping and Forming of Composite Materials, MS&T, Salt Lake City (2016)
7. Carpenter, J.S., et al.: Bulk texture evolution of nanolamellar Zr–Nb composites processed via accumulative roll bonding. Acta Mater. **92**, 97–108 (2015)
8. Liu, C.Y., et al.: Microstructures and mechanical properties of Al/Zn composites prepared by accumulative roll bonding and heat treatment. Mater. Sci. Eng. A. **580**, 36–40 (2013)
9. Ma, M., Huo, P., Liu, W.C., Wang, G.J., Wang, D.M.: Microstructure and mechanical properties of Al/Ti/Al laminated composites prepared by roll bonding. Mater. Sci. Eng. A. **636**, 301–310 (2015)
10. Motevalli, P.D., Eghbali, B.: Microstructure and mechanical properties of tri-metal Al/Ti/Mg laminated composite processed by accumulative roll bonding. Mater. Sci. Eng. A. **628**, 135–142 (2015)
11. Mahdavian, M.M., Ghalandari, L., Reihanian, M.: Accumulative roll bonding of multilayered Cu/Zn/Al: an evaluation of microstructure and mechanical properties. Mater. Sci. Eng. A. **579**, 99–107 (2013)
12. Jafarian, H., Habibi-Livar, J., Razavi, S.H.: Microstructure evolution and mechanical properties in ultrafine grained Al/TiC composite fabricated by accumulative roll bonding. Compos. Part B. **77**, 84–92 (2015)
13. Liu, C.Y., et al.: Evaluation of mechanical properties of 1060-Al reinforced with WC particles via warm accumulative roll bonding process. Mater. Des. **43**, 367–372 (2013)
14. Ahmadi, A., Toroghinejad, M.R., Najafizadeh, A.: Evaluation of microstructure and mechanical properties of Al/Al2O3/SiC hybrid composite fabricated by accumulative roll bonding process. Mater. Des. **53**, 13–19 (2014)
15. Beni, H.A., et al.: Investigation of grain refinement in Al/Al2O3/B4C nano-composite produced by ARB. Compos. Part B. **58**, 438–442 (2014)
16. Chaudhari, G.P., Acoff, V.: Cold roll bonding of multi-layered bi-metal laminate composites composites. Compos. Sci. Technol. **69**, 1667–1675 (2009)
17. Tsuji, N., et al.: ARB (accumulative roll-bonding) and other new techniques to produce bulk ultrafine grained materials. Adv. Eng. Mater. **5**, 338–344 (2003)

Chapter 11
Vibration Characteristics of Unit Cell Structures Fabricated by Multi-Material Additive Manufacturing

Toshitake Tateno and Shogo Nishie

Abstract Natural vibration frequency of objects fabricated by Additive Manufacturing (AM) is experimentally investigated. In this paper, several beam structures consisting of square unit cells are fabricated by material extrusion type AM, which is usually called Fused Deposition Modeling (FDM). The target object is oscillated by a magnetic motor and the amplitude of vibration is measured with a laser vibrometer. The effects of deposition direction and material composition on the primary natural frequency are discussed. These factors can be controlled by using AM. The experimental results showed that the frequency obviously depends on both deposition direction and material composition. The possibility of a vibration design method in product design was indicated.

Keywords Additive Manufacturing • Material extrusion • Vibration • Natural frequency • Deposition direction

11.1 Introduction

Additive Manufacturing (AM) is expected to be used for practical mechanical parts and products. Since an object in AM process is fabricated by depositing materials layer by layer, the fabricated object has anisotropic mechanical properties affected by the deposition direction. There are some studies discussing the effect of deposition direction on mechanical strength [1, 2], but studies on natural frequency are not enough.

Especially, unit cell structures receive significant effects because they consist of fine and complex shapes, in which geometrical errors in deposition become relatively large. Unit cell structures are often adopted as inner structures of parts fabricated by AM because it can be built by AM easily. Such structures can realize a light weight body with keeping strength and stiffness. Since designers are often required to reduce the weight of parts, the unit cell structure is a hope to realize innovative mechanical products.

Additionally, some recent AM machines can build objects by using multi-material AM, in which several kinds of material are located at prescribed positions precisely. Both anisotropy by deposition direction and fabrication with multi-material are features of AM. They have possibility to create innovative products that cannot be achieved with conventional manufacturing.

Material extrusion is a type of AM and is widely used from aerospace industry to DIY at home. Considering practical use of AM parts, it is important to investigate the mechanical properties.

11.2 Experimental Method

Figure 11.1 shows a photograph of a fabricated object example. Arrows are additionally drawn to indicate the deposition directions named as vertical, flat, and horizontal settings. Horizontal and flat settings belong to a laid orientation in fabrication, and vertical condition is a standing orientation. The target object is a beam structure consisting of a weight holder shape and five unit cells. The unit cell is a cubic hollow shape with two open faces. As for the multi-material object, the side plates connecting upper and lower plates are fabricated with a soft material. The target object was built by a material extrusion type AM machine (Flashforge, Dreamer), which has double head extruder. Even when a single-material object is fabricated, two kinds of materials are desired because one material is used for a target object and another is used for support.

T. Tateno (✉) • S. Nishie
Department of Mechanical Engineering Informatics, Meiji University, 1-1-1 Higashimita, Tama-ku, Kawasaki, Kanagawa 214-8571, Japan
e-mail: tateno@meiji.ac.jp

© The Society for Experimental Mechanics, Inc. 2018

J. Wang et al. (eds.), *Mechanics of Additive and Advanced Manufacturing, Volume 9*,
Conference Proceedings of the Society for Experimental Mechanics Series, DOI 10.1007/978-3-319-62834-9_11

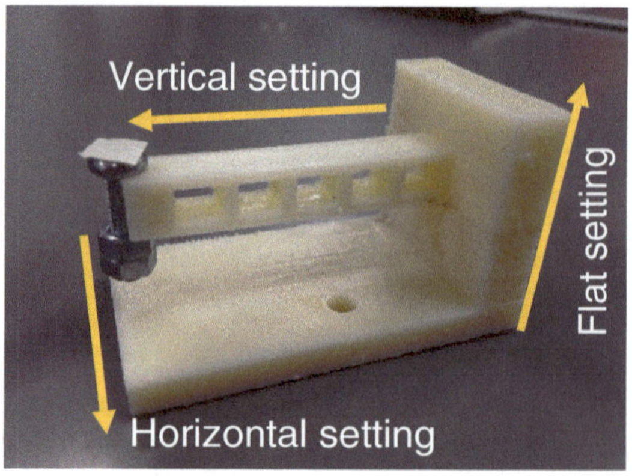

Fig. 11.1 Test object shape and deposition directions for each fabrication setting

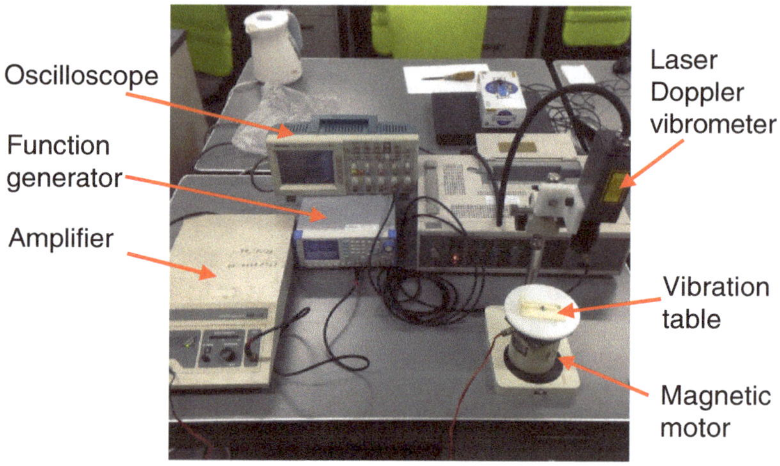

Fig. 11.2 Experimental apparatus for measuring the primary natural frequency with forced vibration

In this study, single-material objects are fabricated with ABS for a target object and HIPS for support, which is soluble in limonene. Multi-material objects are built as horizontal setting with ABS as a hard material and TPU as a soft material and support, which is cut out with a hot cutter after AM process. To measure the vibration clearly, a weight (5.5 g) is set on the weight holder at the end of the beam so that the sympathetic vibration becomes large.

Figure 11.2 shows the experimental apparatus to measure the vibration in forced oscillation. The target object including an angular fixture is set on a magnetic motor. The amplitude and frequency can be controlled with an amplifier and a function generator, respectively. The vibrations on the target object and the vibration table in the vertical direction are measured with a laser Doppler vibrometer. By investigating amplitude ratio between these vibrations, the primary natural frequency is found.

As a reference of natural frequencies, FEM analysis is executed by using 3D-CAD model and ABS material properties.

11.3 Results and Discussion

Figure 11.3 shows experimental and simulation results of the primary natural frequency. Comparing vertical, flat, and horizontal settings, the frequencies were varied due to the deposition direction even though the same shape and material are used. The frequency of simulation result was lower than that of experimental results. One of the reason is that the AM process generates some material gap inside the object and shape errors on the surface, which affect stiffness of the beam. With regard to the multi-material object, the frequency was significantly lower than that of the single-material object in horizontal setting. Since some hard material was replaced with soft material, lowering of frequency is a natural result. However, it showed that a little changing of material can generate large change of vibration. The amount of change is not investigated but it must be replicable. By compiling experimental tests, the amount will become to estimate.

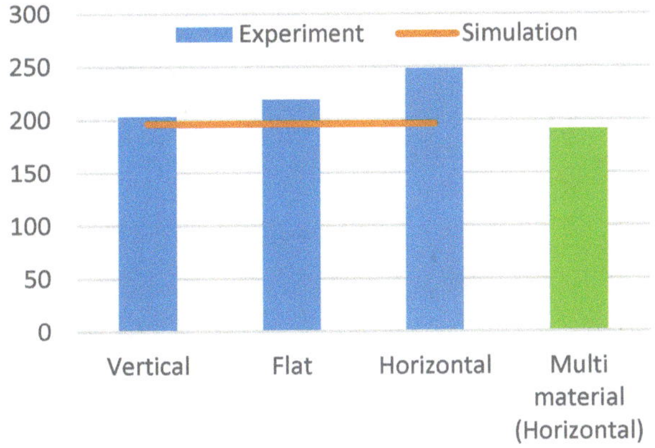

Fig. 11.3 The primary natural frequencies of single-material objects and a multi-material object

11.4 Conclusion

In this study, vibration properties of objects fabricated by AM were explored. The vibration experiments for measuring the primary natural frequency were conducted. The experimental results showed that the natural frequency depends on deposition direction in AM. Additionally, the primary natural frequency of multi-material object, of which some portion is made with soft material, was measured with the same method. The experimental result had a large difference from that of single-material objects. Both the anisotropy by deposition direction and the multi-material fabrication are features of AM. If the amount of variation can be estimated, it is possible to be a vibration design method in product design.

Reference

1. Park, S.-i., Rosen, D.W.: Quantifying effects of material extrusion additive manufacturing process on mechanical properties of lattice structures using as-fabricated voxel modeling. Addit. Manuf. **89**, 1–9 (2016)
2. Tateno, T., Yaguchi, Y., Fujitsuka, M.: Study on parallel fabrication in additive manufacturing (connector design considering anisotoropic strength by deposition direction). J. Adv. Mech. Des. Syst. Manuf. **10**(7), Paper No.15–00716 (2016)

Chapter 12
Defects, Process Parameters and Signatures for Online Monitoring and Control in Powder-Based Additive Manufacturing

Ehsan Malekipour and Hazim El-Mounayri

Abstract Additive Manufacturing (AM) is a process that is based on manufacturing parts layer by layer in order to avoid any geometric limitation in terms of creating the desired design. In the early stages of AM development, the goal was just creating some prototypes to decrease the time of manufacturing assessment. But with metal-based AM, it is now possible to produce end-use parts. In powder-based AM, a designed part can be produced directly from the STL file (Standard Tessellation Language/ stereolithography) layer by layer by exerting a laser beam on a layer of powder with thickness between 20 μm and 100 μm to create a section of the part. The Achilles' heel of this process is generation of some defects, which weaken the mechanical properties and in some cases, these defects may even lead to part failure under manufacturing. This prevents metal-based AM technology from spreading widely while limiting the repeatability and precision of the process. Online monitoring (OM) and intelligent control, which has been investigated prevalently in contemporary research, presents a feasible solution to the aformentioned issues, insofar as it monitors the generated defects during the process and eliminates them in real-time. In this regard, this paper reveals the most frequent and traceable defects which significantly affect quality matrices of the produced part in powder-based AM, predominately focusing on the Selective Laser Sintering (SLS) process. These defects are classified into "Geometry and Dimensions," "Surface Quality (Finishing)," "Microstructure" and the defects leading to "Weak Mechanical Properties." In addition, we introduce and classify the most important parameters, which can be monitored and controlled to avoid those defects. Furthermore, these parameters may be employed in some error handling strategies to remove the generated defects. We also introduce some signatures that can be monitored for adjusting the parameters into their optimum value instead of monitoring the defects directly.

Keywords Additive manufacturing • Powder-based additive manufacturing • Defects and process parameters • Online monitoring and control • Intelligent control

12.1 Introduction

Additive Manufacturing (AM) is a process which is based on manufacturing parts layer by layer in order to avoid any geometric limitation in terms of creating the desired design. AM was started by producing parts using the method of stereolithography. In the early stages of AM development, the goal of AM was just creating some prototypes to decrease the time of manufacturing assessment. But with metal-based AM [1–7] it is now possible to produce end-used parts. In Selective Laser Sintering (SLS), a designed part can be produced directly from the STL file layer by layer by exerting a laser beam on a layer of powder with thickness between 20 μm and 100 μm to sinter the powder particles together to create a section of the part, the thickness of layer. The Achilles' heel of this process is generating some defects during the process which lead to the weakening of the mechanical properties for some products. In some cases these defects lead to part failure under manufacturing or during post-processing/operation. This prevents the technology from spreading widely and limits the repeatability and precision of the process. In following, we classified the defects, signatures and process parameters which can be employed in online monitoring and control.

Standard Tessellation Language (stereolithography).

E. Malekipour (✉) • H. El-Mounayri
Center for Additive Manufacturing Research at IUPUI (CAMRI), Purdue School of Engineering and Technology, Indianapolis, IN, USA
e-mail: emalekip@purdue.edu; helmouna@iupui.edu

© The Society for Experimental Mechanics, Inc. 2018

J. Wang et al. (eds.), *Mechanics of Additive and Advanced Manufacturing, Volume 9*,
Conference Proceedings of the Society for Experimental Mechanics Series, DOI 10.1007/978-3-319-62834-9_12

12.2 Classification of Defects for Online Monitoring and Control

There are two main approaches to control the defects during AM process, which are described below.

The first approach uses analytical models, in order to predict the values of the process parameters [8–14]. However, there is a lack of mathematical and statistical models of and algorithms for the AM process to accurately predict process parameters in order to avoid failures, improve the part quality and produce a perfect product. These parameters are employed to account for the process specifications such as the material, ambient temperature, geometry, required speed of manufacturing, scan pattern, etc. To date, all of the conducted investigations are based on simulation and physics-based Finite Element Analysis (FEA) that are complex and, introduce high computational burden [15, 16]. Efficient analytical and data driven models which are capable of processing large data streams are strongly needed for real-time control. This limitation is due to the complex nature of the sintering process due to the change in material properties as a result of increased temperature, in the plurality of contributing parameters and process parameters, and the lack of understanding of physical and chemical reactions between powders during the process. These problems encouraged scholars to consider the second approach.

The second approach is Online Monitoring (OM). The most significant advantages of this approach are: (1) It can be implemented without a full modeling of the physics of sintering phenomena; (2) It can be employed to eliminate the generated defects precisely. There is a good volume of literature on the different types of defects to improve the quality of the final produced part by adjusting the contributing parameters in the suitable range. However, there is a lack of effective/systematic classification of the defects which significantly affects quality matrices of the produced part and their contributing parameters. Also, the correlation between those defects and contributing parameters. In addition, no focus on the most important parameters which can be monitored and controlled to avoid those defects; the literature, however, shows some primary grouping of parameters [7].

As it is shown in Fig. 12.1, we classified 13 different defects related to "Geometry and Dimensions", "Surface Quality (Finishing)", "Microstructure" and the defects behind "Weak Mechanical Properties".

From online monitoring point of view, we can reorganize all the mentioned defects under a quintuple set of manufacturing features revealed here. If we can monitor and control this set of features, achieving in flawless part will be applicable [17–24].

1. *The homogeneous deposition of the powder*
   As it is already mentioned uniformly Powder deposition, smoothness of every deposited powder layer and uniform spreading are very curtail for surface quality.
2. *Thermal characteristics of the layer under fabrication*
   Temperature characteristics such as homogeneous temperature distribution, heat accumulation in different zones, etc. is an excellent proxy [25] in order to attain the best possible homogenous microstructure which directly effect on mechanical properties as one of the most important demands in SLS manufacturing.
3. *Surface quality related defects for a single layer*
   It includes a vast number of defects such as cracks, holes, etc. It is noteworthy that some of the defects such as porosity needs to be checked after every specific number of layers [26–28].
4. *Improper part geometry and dimension inaccuracy*:
   This is the most important category regarding repeatability. As explained in Sect. 12.1 defects such as shrinkage leads to some inaccuracy geometrically and dimensionally [26].

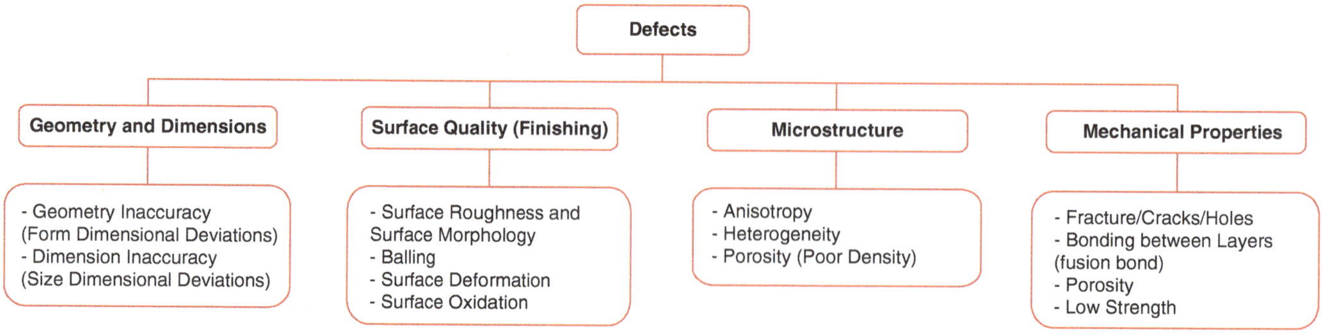

Fig. 12.1 Common defects in SLS process

5. *Poor bonding between the layers:*
   It is very effective on the mechanical properties and mainly affected by energy density, penetration depth and layer thickness [28, 29].

Also, it has been turned out that all of these defects/features can be handled by just a small set of controllable parameters. In other words, by controlling those few parameters, vendors can prevent defects in some cases or eliminate them in some other cases, after detecting, and as a result, reach the best possible quality for the fabricated part [18, 30–32]. These parameters are derived, classified and explained in the following section. It also should be noticed that, all of these fields are able to be monitored by current instruments and there are numerous approaches which are introduced in literature to monitor some of these defects/features [33–44].

12.3 Classification of Process Parameters for Online Monitoring and Control

Nowadays, the parts produced in SLS are used as end-used products. Therefore, producing a part with perfect homogeneous microstructure and exclusive defects is considered as ultimate objective of SLS process. In this regard, there are some barriers, challenges and gaps including powder properties knowledge, chain capabilities measurement science, standardization, monitoring defects and control manufacturing process. "Energetics Incorporated" addressed these items and prepared a road map for NIST to show the gaps needed to be filled [45]. Online monitoring and process control as an ongoing discussion can help to develop this process significantly. In this context, there are various defects, signatures and parameters affecting the quality of a printed part and its microstructure which can be controlled by aligning process parameters with optimized values through online monitoring. Hence, it is crucial to utilize some techniques to monitor and control process parameters and probably, desired signatures constantly in order to avoid the defects. To further establish a thorough procedure for the monitoring and controller strategy, the process parameters can be classified into three categories as shown in Fig. 12.2. As this figure depicts, the first category is "pre-processed parameters" which should be specified before starting the process. This category is divided into two subcategories itself. The first one, the "pre-defined parameters", includes the items which must be chosen or set before the manufacturing process. There is an optimum value or a best state to be chosen for each of them. It is noteworthy that these parameters are constant for any SLS process regardless of the manufactured part. Consequently, vendors will not need to monitor or control them. This category mostly include powder specifications and machine specifications.

The second subcategory, the "Pre-defined parameters that need to be monitored", includes the parameters inside the chamber as well as machine error parameters. These parameters must be specified before starting the manufacturing process as well, but they can be changed during the process; however, these changes are undesirable and thus, need to be monitored and kept constant. It is important to note that unlike the first group of parameters, they may have a different optimum value or state, based on the chosen material. To sum up, we would need to monitor them in situ to keep them fix but we would not need to control them to alter their values during the process.

The second category of parameters are the most important parameters for online monitoring in SLS process. This category which is called "controllable parameters" includes process parameters (laser specifications and scan strategy) and manufacturing specifications. These parameters need to be monitored and altered in situ based on: (1) the feedback that controller received from the sensors to adjust these parameters with the benchmarks information; (2) manufacturing strategy such as laser pattern; and (3) the chosen manufacturing preference such as time of manufacturing, desired mechanical properties, density, etc. It should be noted that currently all of these parameters are pre-defined and constant during the manufacturing process; however, the manufacturers can assign the correct value or choose the suitable state of process parameters based on the mentioned items. There are two main reasons why these parameters should be monitored and controlled. The first reason is to avoid as much defects as possible and reach the best possible end-used part. For instance, one recommended potential application is altering the pace of manufacturing by using thicker layer in parts with perpendicular walls to the base plate. In this case we would need to adjust other parameters such as laser power or laser speed to keep strong bonding between layers. It should be noted that all of these parameters can be easily measured and controlled by the machine.

The second reason is to eliminate the defects after their detection when possible. Actually, defects are created due to some manufacturing uncertainty such as the imprecision of the process, stochastic spreading of powders, the inability to predict and choose the most suitable process parameters, etc. The reason is that, there is no analytical model that can be used reliably for prediction and explain the process completely [25].

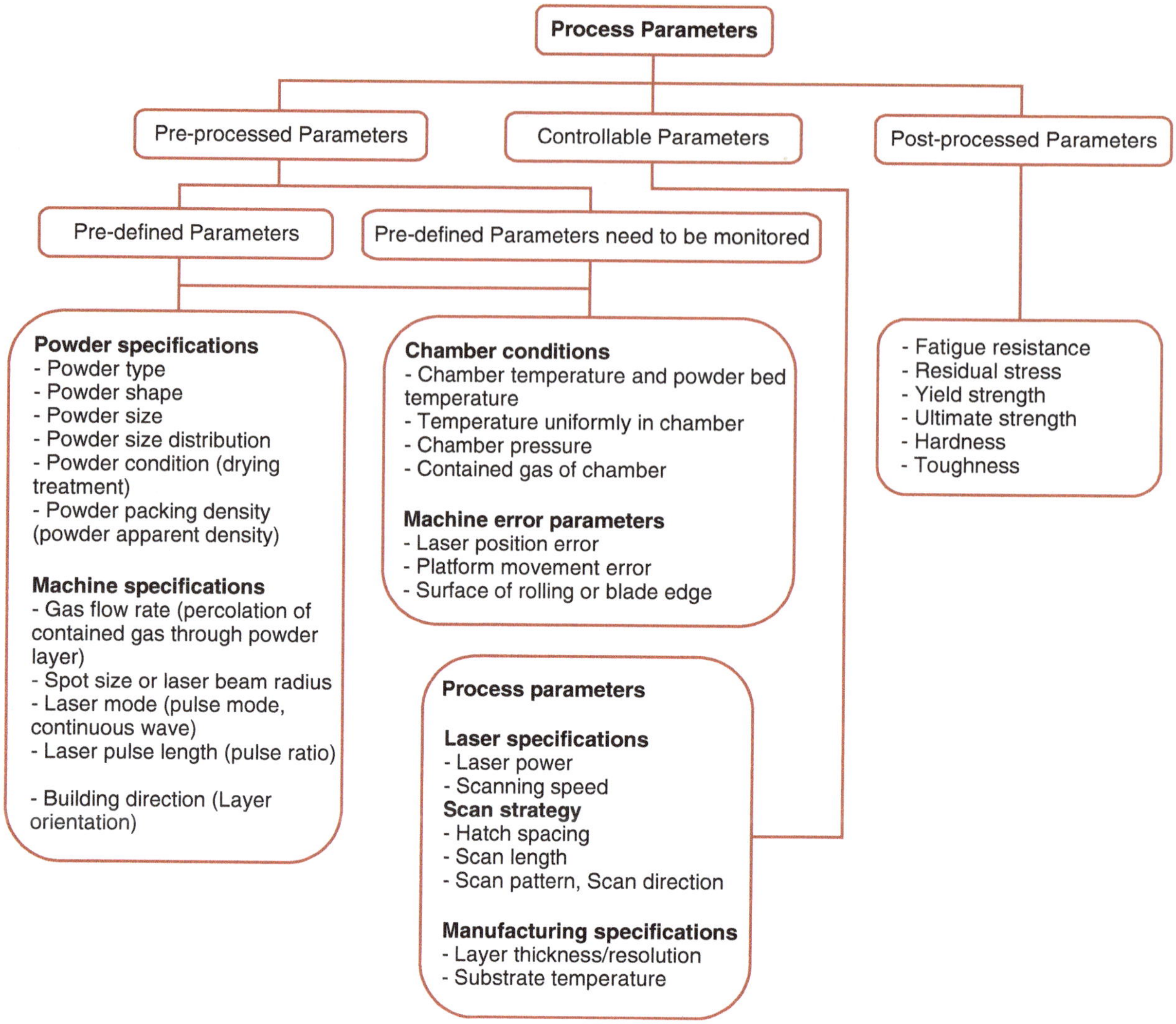

Fig. 12.2 Process parameters classification

The final category is post-process parameters witch are not possible to be monitored during the process. These parameters such as yield strength, ultimate strength, etc. showing the mechanical properties of the manufactured part, are among the most desired demands of manufacturing and they are affected by microstructures defects and grain structures, For example, micro crack can lead to stress concentration which reduces life time in fatigue condition [46]. We can improve the mechanical properties by controlling the contributing parameters affecting microstructure. At the very end of the process, these parameters show the quality of the manufacturing process and the success rate of online monitoring approach.

Table 12.1 shows how contributing parameters affect the defects or process features (effected items) in powder-based AM process. As it reveals, there are only five parameters which need to be monitored and kept fixed (second category) and 10 parameters which need to be monitored and controlled (third category). These parameters are recommended for online monitoring and control of powder-based process specifically for SLS process.

12.4 Using of Signatures in Online Process Control

Another way to perform monitoring in SLS or other AM processes is to monitor and measure some signatures instead of monitoring defects and parameters directly. Signatures are defined as some manufacturing specifications or combination of some parameters which may be utilized to adjust the controllable parameters to avoid the defects. We can divide Signatures

Table 12.1 Parameter classification using for online monitoring in SLS process

No.	Process parameter	Effected items
First category: predefined parameters		
1	Powder shape	Surface contamination, flowability of bulk of powder (smooth deposition), powder packing density and powder apparent density, porosity, powder fluidity
2	Powder size	Flowability of bulk of powder, balling, powder packing density and powder apparent density, material strength, oxidation, layer thickness
3	Powder size distribution	Powder packing density and powder apparent density
4	Powder condition/ drying treatment	Balling, heterogeneity, porosity, oxidation
5	Powder packing density/ powder apparent density	Porosity/ density
6	Gas flow (rate and direction)	Dimensional tolerance, bonding between layers
7	Building direction (layer orientation)	Dimension inaccuracy, anisotropy, porosity, mechanical properties
8	Spot diameter and beam radius	Dimension inaccuracy, surface roughness, porosity (penetration depth), energy density and energy absorption
9	Laser mode (pulse mode, continuous wave)	Porosity (density), balling (wetting angle)
10	Laser pulse length/ pulse ratio	Surface roughness, bonding between layers
Second category: predefined parameters that need to be monitored		
11	Chamber temperature and powder bed temperature (workpiece temperature)	Shrinkage, heterogeneity, surface deformation, absorptivity
12	Gas flow rate	Dimensional inaccuracy, porosity, mechanical properties, bonding between layers
13	Contained gas of chamber	Balling, oxidation
14	Percolation of contained gas through powder layer	Surface oxidation
Third category: parameters should be monitored and controlled		
15	Laser power	Shrinkage, balling (energy density), warping, layer distortion, porosity, melt pool size and morphology, penetration depth
16	Scanning speed	Shrinkage, surface roughness, balling (energy density, poor wetting), layer distortion, porosity, melt pool size and morphology, fracture/cracks/holes
17	Laser positioning error	Geometry inaccuracy
18	Hatch spacing	Shrinkage, surface roughness, energy density, heterogeneity, porosity, melt pool size and morphology, dimension inaccuracy (microstructural waviness), surface deformation (heat accumulation)
19	Scan length	Shrinkage, warping, fracture/cracks/holes
20	Scan pattern, scan direction	Surface roughness, porosity/ density, heterogeneity, anisotropy
21	Platform movement error	Geometry inaccuracy
22	Layer thickness/ resolution	Staircase effect, shrinkage, energy density, porosity/density, bonding between layers/ penetration depth, construction speed
23	Blade edge or surface of rolling	Pit on the surface, powder deposition/ distribution
24	Substrate temperature	Balling, distortion and warpage, poor wetting, melt pool shape

into two different types: The first type can be defined as "Manufacture Signature". This type of signatures are the manufacturing specifications which are affected clearly by the controllable parameters and hence, this correlation may be utilized to adjust the controllable parameters. It should be noted that all of these Manufacture Signature relate to melting phenomena and melt poll specification [19, 47–50] Thus, they are suitable to be used in SLM process. The second type of signature can be defined as "Cumulative Signature" which consists of number of controllable parameters which are able to be set according to some prepared benchmark. These parameters can be used for both SLM and SLS process [18, 34, 51]. The melt pool specifications, Marangoni convection and a feature set of laser scan are recommended in literature (or here for the first time) to be used as signatures in monitoring approaches. The detailed explanation about the correlation between these items and their contributing parameters was mentioned in Sect. 12.2.

12.4.1 Manufacturing Signature (for SLM Process)

12.4.1.1 Melt Pool Morphology and Melt Pool Dimensions

Melt pool morphology (and as a result melt pool dimension including melt pool depth) has a strong effect on the bonding between scan tracks and as a result, inter-run porosity and final density, balling, Heat Affected Zone (HAZ), penetration depth and porosity [52–54]. In HAZ area the dimensions of melt pool grow because of lack of heat conductivity and thus, the surface quality declines significantly. As it is revealed in reference [34], generating support structure can reinforce these areas while improve heat conductivity which leads to better surface quality. The parameters contributed in melt pool morphology are scan speed, laser power, spot size, the type of powder, overlapping ratio and the number of laser scan. We can also map the melt pool dimensions with the generated microstructure to indirectly control solidification microstructure [55, 56].

12.4.1.2 Melt Pool Temperature and Solidification Rate

Melt pool temperature, solidification rate, temperature gradient and scan speed are closely related to each other and significantly affect solidification microstructure, homogeneity, and type of grain structure (cellular or dendritic). The governing equation is

$$\frac{\mathrm{G}}{\mathrm{R}} = \frac{T_p}{\mathrm{u}x_1} \tag{12.1}$$

Where T_p is melt pool temperature, u is scan speed, x_1 is distance between the heat source and the rear of the weld pool, G is temperature gradient and R is solidification velocity. Furthermore, melt pool temperature affects melt pool depth whose correlation is shown Fig. 12.3; the solidification rate affects porosity (with entrapping gas), shrinkage, balling and melt pool geometry.

12.4.1.3 Marangoni Convection (MC)

Altering the temperature coefficient of surface tension (due to variation of the solute oxygen) and generating large thermal gradient between the center and edge of the melt pool (due to the use of laser Gaussian distribution) leads to surface tension gradient. This tension gradient triggers fluid flow in melt pool causing changes in temperature gradient. This phenomena, called MC, changes the melt penetration, generates deep narrow tracks (leading to inter-run porosity), changes bonding between layers, the ratio of depth/width of melt pool and the melt pool morphology (cross section of melt pool). Balling, humping, changes surface morphology, solidification microstructure and warping are other effected items [12, 52, 57].

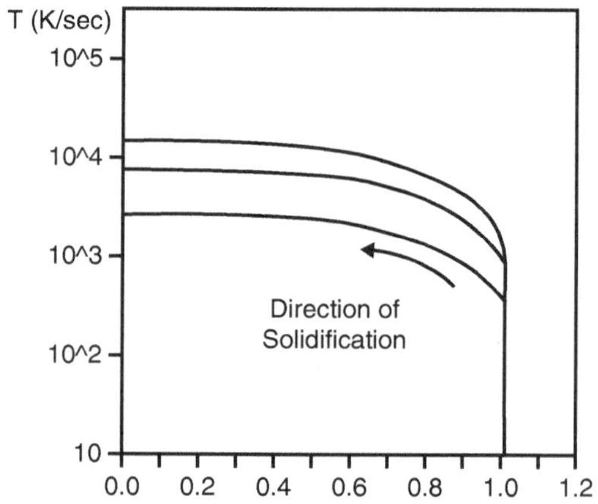

Fig. 12.3 Dependence T vs. dimensionless melt depth at different scan speed

12.4.2 Cumulative Signature (for SLS and SLM)

Overlapping ratio and energy density (energy input) can be introduced as cumulative signatures. Overlapping ratio affects surface roughness, porosity and melt penetration. Energy density however can effect on balling, heterogeneity/homogeneity, density (porosity), melt pool size, and melt pool morphology, wetting angle and solidification.

12.5 Conclusion and Future Works

This paper reveals the most common defects in SLS process and also, classifies the defects addressed by literatures as the most prevalent defects in five categories to use for monitoring purposes. It also classifies process parameters in pre-defined, controllable and post-process ones and recommends to set the controllable parameters on their optimized values in situ to avoid the defects.

This paper also elaborates on the concept of signatures and introduces some potential signatures which can be employed in order to control the quality of the fabricated parts indirectly.

It is noteworthy that however literatures introduced a set of efficient approaches to monitor some of these defects/ signatures, there is no integrated approach to cover all the defects and also, there is no impressive approaches to eliminate the detected defects efficiently.

References

1. Frazier, W.E.: Metal additive manufacturing: a review. J. Mater. Eng. Perform. **23**, 1917–1928 (2014)
2. Ruan, J., Sparks, T. E., Fan, Z., Stroble, J. K., Panackal, A., Liou, F.: A review of layer based manufacturing processes for metals. In: 17th Solid Freeform Fabrication Symposium, Austin, pp. 233–245 (2006)
3. Wong, K. V., Hernandez, A.: A review of additive manufacturing. ISRN Mech. Eng. **2012**, 1–10 (2012)
4. Kulkarni, P., Marsan, A., Dutta, D.: A review of process planning techniques in layered manufacturing. Rapid Prototyp. J. **6**, 18–35 (2000)
5. Gu, D., Meiners, W., Wissenbach, K., Poprawe, R.: Laser additive manufacturing of metallic components: materials, processes and mechanisms. Int. Mater. Rev. **57**, 133–164 (2012)
6. Bhavar, V., Kattire, P., Patil, V., Khot, S., Gujar, K., Singh, R.: A review on powder bed fusion technology of metal additive manufacturing. In: International Conference & Exhibition on Additive Manufacturing Technologies, Bangalore (2015)
7. Gong, X., Anderson, T., Chou, K.: Review on powder-based electron beam additive manufacturing technology. In: ASME/ISCIE 2012 International Symposium on Flexible Automation, pp. 507–515 (2012)
8. Flood, A., Liou, F.: Modeling of powder bed processing–a review. In: International Solid Freeform Fabrication Symposium (2015)
9. Ganeriwala, R., Zohdi, T.I.: Multiphysics modeling and simulation of selective laser sintering manufacturing processes. Procedia CIRP. **14**, 299–304 (2014)
10. Zohdi, T.I.: Rapid simulation of laser processing of discrete particulate materials. Arch. Comput. Meth. Eng. **20**, 309–325 (2013)
11. Xiao, B., Zhang, Y.: Laser sintering of metal powders on top of sintered layers under multiple-line laser scanning. J. Phys. D. Appl. Phys. **40**, 6725 (2007)
12. Xiao, B., Zhang, Y.: Marangoni and buoyancy effects on direct metal laser sintering with a moving laser beam. Numer. Heat Transfer, Part A Appl. **51**, 715–733 (2007)
13. Zeng, K., Pal, D., Stucker, B.: A review of thermal analysis methods in laser sintering and selective laser melting. In: Proceedings of Solid Freeform Fabrication Symposium, Austin (2012)
14. Attar, E.: Simulation of Selective Electron Beam Melting Processes. Dr.-Ing., University of Erlangen, Nuremberg (2011)
15. N'Dri, N., Mindt, H. W., Shula, B., Megahed, M., Peralta, A., Kantzos, P., et al.: DMLS process modelling & validation. In: TMS2015 Supplemental Proceedings, pp. 389–396 (2015)
16. King, W., Anderson, A., Ferencz, R., Hodge, N., Kamath, C., Khairallah, S.: Overview of modelling and simulation of metal powder bed fusion process at Lawrence Livermore National Laboratory. Mater. Sci. Technol. **31**, 957–968 (2015)
17. Spierings, A., Herres, N., Levy, G.: Influence of the particle size distribution on surface quality and mechanical properties in AM steel parts. Rapid Prototyp. J. **17**, 195–202 (2011)
18. Craeghs, T., Bechmann, F., Berumen, S., Kruth, J.-P.: Feedback control of layerwise laser melting using optical sensors. Phys. Procedia. **5**, 505–514 (2010)
19. Craeghs, T., Clijsters, S., Yasa, E., Kruth, J.-P.: Online quality control of selective laser melting. In: Proceedings of the Solid Freeform Fabrication Symposium, pp. 212–226, Austin (2011)
20. Hu, D., Kovacevic, R.: Sensing, modeling and control for laser-based additive manufacturing. Int. J. Mach. Tools Manuf. **43**, 51–60 (2003)
21. Hu, D., Mei, H., Kovacevic, R.: Improving solid freeform fabrication by laser-based additive manufacturing. Proc. IMechE B J. Eng. Manuf. **216**, 1253–1264 (2002)
22. Mahesh, M., Wong, Y., Fuh, J., Loh, H.: A six-sigma approach for benchmarking of RP&M processes. Int. J. Adv. Manuf. Technol. **31**, 374–387 (2006)

23. Paul, R., Anand, S.: Process energy analysis and optimization in selective laser sintering. J. Manuf. Syst. **31**, 429–437 (2012)
24. Khaing, M., Fuh, J., Lu, L.: Direct metal laser sintering for rapid tooling: processing and characterisation of EOS parts. J. Mater. Process. Technol. **113**, 269–272 (2001)
25. Tapia, G., Elwany, A.: A review on process monitoring and control in metal-based additive manufacturing. J. Manuf. Sci. Eng. **136**, 060801 (2014)
26. Delgado, J., Ciurana, J., Rodríguez, C.A.: Influence of process parameters on part quality and mechanical properties for DMLS and SLM with iron-based materials. Int. J. Adv. Manuf. Technol. **60**, 601–610 (2012)
27. Yasa, E., Deckers, J., Craeghs, T., Badrossamay, M., Kruth, J.-P.: Investigation on occurrence of elevated edges in selective laser melting. In: International Solid Freeform Fabrication Symposium, pp. 673–85, Austin (2009)
28. Meier, H., Haberland, C.: Experimental studies on selective laser melting of metallic parts. Mater. Werkst. **39**, 665–670 (2008)
29. Abd-Elghany, K., Bourell, D.: Property evaluation of 304L stainless steel fabricated by selective laser melting. Rapid Prototyp. J. **18**, 420–428 (2012)
30. Berumen, S., Bechmann, F., Lindner, S., Kruth, J.-P., Craeghs, T.: Quality control of laser-and powder bed-based Additive Manufacturing (AM) technologies. Phys. Procedia. **5**, 617–622 (2010)
31. Llanos, I., Agirre, A., Quintana, I., Urreta, H.: Defect elimination on laser milled geometries by means of error detection and correction. Int. J. Adv. Manuf. Technol. **84**(1–8), 2471–2478 (2016)
32. Verma, A., Tyagi, S., Yang, K.: Modeling and optimization of direct metal laser sintering process. Int. J. Adv. Manuf. Technol. **77**, 847–860 (2015)
33. Kleszczynski, S., Zur Jacobsmühlen, J., Sehrt, J., Witt, G.: Error detection in laser beam melting systems by high resolution imaging. In: Proceedings of the Solid Freeform Fabrication Symposium (2012)
34. Craeghs, T., Clijsters, S., Kruth, J.-P., Bechmann, F., Ebert, M.-C.: Detection of process failures in layerwise laser melting with optical process monitoring. Phys. Procedia. **39**, 753–759 (2012)
35. Chivel, Y., Smurov, I.: On-line temperature monitoring in selective laser sintering/melting. Phys. Procedia. **5**, 515–521 (2010)
36. Lott, P., Schleifenbaum, H., Meiners, W., Wissenbach, K., Hinke, C., Bültmann, J.: Design of an optical system for the in situ process monitoring of selective laser melting (SLM). Phys. Procedia. **12**, 683–690 (2011)
37. Yadroitsev, I., Krakhmalev, P., Yadroitsava, I.: Selective laser melting of Ti6Al4V alloy for biomedical applications: temperature monitoring and microstructural evolution. J. Alloys Compd. **583**, 404–409 (2014)
38. Dinwiddie, R. B., Dehoff, R. R., Lloyd, P. D., Lowe, L. E., Ulrich, J. B.: Thermographic in-situ process monitoring of the electron-beam melting technology used in additive manufacturing. In: SPIE Defense, Security, and Sensing, pp. 87050K–87050K-9 (2013)
39. Price, S., Lydon, J., Cooper, K., Chou, K.: Experimental temperature analysis of powder-based electron beam additive manufacturing. In: Proceedings of the Solid Freeform Fabrication Symposium, pp. 162–173 (2013)
40. Rodriguez, E., Medina, F., Espalin, D., Terrazas, C., Muse, D., Henry, C., et al.: Integration of a thermal imaging feedback control system in electron beam melting. In: Proceedings of the Solid Freeform Fabrication Symposium (2012)
41. Wegner, A., Witt, G.: Process monitoring in laser sintering using thermal imaging. In: SFF Symposium, pp. 8–10, Austin (2011)
42. Krauss, H., Eschey, C., Zaeh, M.: Thermography for monitoring the selective laser melting process. In: Proceedings of the Solid Freeform Fabrication Symposium (2012)
43. Santospirito, S., Słyk, K., Luo, B., Łopatka, R., Gilmour, O., Rudlin, J.: Detection of defects in laser powder deposition (LPD) components by pulsed laser transient thermography. In: SPIE Defense, Security, and Sensing, pp. 87050X–87050X-11 (2013)
44. Price, S., Cooper, K., Chou, K.: Evaluations of temperature measurements by near-infrared thermography in powder-based electron-beam additive manufacturing. In: Proceedings of the Solid Freeform Fabrication Symposium, pp. 761–773, Austin (2012)
45. Kerwien, S., Collings, S., Liou, F., Bytnar, M.: Measurement Science Roadmap for Metal-based Additive Manufacturing, ed: NIST (2013)
46. Ellyin, F.: Fatigue Damage, Crack Growth and Life Prediction. Springer Science & Business Media, London (2012)
47. Aggarangsi, P., Beuth, J. L., Griffith, M.: Melt pool size and stress control for laser-based deposition near a free edge. In: Solid Freeform Fabrication Proceedings, pp. 196–207 (2003)
48. Aggarangsi, P., Beuth, J. L., Gill, D. D.: Transient changes in melt pool size in laser additive manufacturing processes. In: Solid Freeform Fabrication Proceedings, University of Texas, Solid Freeform Fabrication Symposium, pp. 2–4, Austin (Aug, 2004)
49. Birnbaum, A., Aggarangsi, P., Beuth, J.: Process scaling and transient melt pool size control in laser-based additive manufacturing processes. In: Solid Freeform Fabrication Proceedings, pp. 328–339, Austin (2003)
50. Boddu, M. R., Landers, R. G., Liou, F. W.: Control of laser cladding for rapid prototyping—A review. In: Proceedings of the Solid Freeform Fabrication Symposium, pp. 6–8, The University of Texas at Austin, Austin (2001)
51. Bi, G., Gasser, A., Wissenbach, K., Drenker, A., Poprawe, R.: Characterization of the process control for the direct laser metallic powder deposition. Surf. Coat. Technol. **201**, 2676–2683 (2006)
52. Hauser, C.: Selective Laser Sintering of a Stainless Steel Powder. University of Leeds, Leeds (2003)
53. Gong, H., Gu, H., Zeng, K., Dilip, J., Pal, D., Stucker, B., et al.: Melt pool characterization for selective laser melting of Ti-6Al-4V pre-alloyed powder. In: Solid Freeform Fabrication Proceedings, Austin (2014)
54. Scharowsky, T., Osmanlic, F., Singer, R., Körner, C.: Melt pool dynamics during selective electron beam melting. Appl. Phys. A. **114**, 1303–1307 (2014)
55. Cheng, B., Chou, K.: Melt pool geometry simulations for powder-based electron beam additive manufacturing. In: 24th Annual International Solid Freeform Fabrication Symposium-An Additive Manufacturing Conference, Austin (2013)
56. Gockel, J., Beuth, J., Taminger, K.: Integrated control of solidification microstructure and melt pool dimensions in electron beam wire feed additive manufacturing of Ti-6Al-4V. Addit. Manuf. **1**, 119–126 (2014)
57. Gu, D., Shen, Y.: Effects of processing parameters on consolidation and microstructure of W–Cu components by DMLS. J. Alloys Compd. **473**, 107–115 (2009)

Chapter 13
The Effect of the 3-D Printing Process on the Mechanical Properties of Materials

Bobby Tang Dan, Daniel Robert Khodos, Oliver Khairallah, Richi Ramlal, and Yougashwar Budhoo

Abstract Tensile tests were conducted following the ASTM Standard D3039 for different specimens produced using additive manufacturing. In the expansion of the 3D printing market, different methods of three-dimensional printing and materials have been developed. Different specimens ranging in materials and infill were printed with a consumer grade PLA FDM printer without heated bed, a consumer grade SLA printer, and an industrial manufacturing grade ABS FDM printer with heated bed. The specimens included Polylactic Acid (PLA), Tough PLA, Acrylonitrile Butadiene Styrene (ABS), and black resin. The PLA and tough PLA specimens were printed at infill intervals of 20% and at 0/90 and $\pm 45^{\circ}$ with respect to the longitudinal axis. The ABS material was also printed at the same angles with respect to the longitudinal axis at solid (SO), sparse double density (DD), sparse high density (HD), and sparse low density (LO) infill values. The consumer grade SLA printer was used to print specimens using black resin and was divided into two sets. The first set includes specimens that were cured under an ultraviolet light while the second set contained those that were not cured after printing. The overall result showed that the infill and print orientation of $\pm 45^{\circ}$ performed better for the PLA and Tough PLA material. However, the orientation did not influence the ABS material which may have been the result of using an industrial grade 3D printer. In resins, the increase in resolution resulted in a higher ultimate stress while curing normalized the ductility of the material.

Keywords Tensile • Additive manufacturing • PLA • ABS • Resin

13.1 Introduction

The tensile test is one of the most fundamental test used in obtaining the mechanical properties of a material. Additive manufacturing (3D Printing) is the process of producing three dimensional solid objects from a digital file. One method of additive manufacturing, the object is created by laying down successive layers of material until the object is created. Popular methods of 3D printing include Selective Laser Sintering (SLS), Fused Deposition Modeling (FDM), and Stereo lithography (SLA). The SLS method involves the use of a high-power laser to fuse small particles of material such as plastic, metal, and ceramic into the desired three-dimensional object. Fused Deposition Modeling uses a plastic filament or metal wire that is supplied into an extrusion nozzle. The extrusion nozzle heats the material supplied and moves in horizontal and vertical directions with a numerically controlled mechanism while controlling the flow of the material exiting the nozzle. This method of 3D printing is seen in products in many companies. The process of Stereo lithography involves a container of ultraviolet curable photopolymer resin and an ultraviolet laser to build the object's layers one at a time. This process is seen in the FormLabs 2 resin printer. With the expansion and growth of different printing processes and methods, there have been very little research on the effect of the printing process on the mechanical properties of the materials available on the market. This paper looks at the FDM and SLA methods of printing and their effect on tensile properties of the materials printed. Applications of this work can be model prototyping and material selection in production.

B. Tang Dan (✉) • D.R. Khodos • O. Khairallah • R. Ramlal • Y. Budhoo
Department of Engineering and Technology, Vaughn College of Aeronautics and Technology,
8601 23rd Avenue Flushing, New York, NY 11369, USA
e-mail: bobby.tang@vaughn.edu

© The Society for Experimental Mechanics, Inc. 2018
J. Wang et al. (eds.), *Mechanics of Additive and Advanced Manufacturing, Volume 9*,
Conference Proceedings of the Society for Experimental Mechanics Series, DOI 10.1007/978-3-319-62834-9_13

13.2 Experimental Procedure

13.2.1 Materials

The materials studied in this project were 6.35 mm × 25.4 mm × 254 mm specimens of Polylactic Acid (PLA), Tough PLA, Acrylonitrile Butadiene Styrene, (ABS), and Black Resin. The PLA and tough PLA specimens were printed at infill intervals of 20% and at 0/90° and ±45° with respect to the longitudinal axis. The ABS material was also printed at the same angles with respect to the longitudinal axis at solid (SO), sparse double density (DD), sparse high density (HD), and sparse low density (LO) infill values. The consumer grade SLA printer was used to print specimens using black resin at 0.1 and 0.025-millimeter resolution. The black resin was divided into specimens that were cured under an ultraviolet light and specimens that were not cured. A minimum of four tests were conducted for each specimen type that was chosen.

13.2.2 3D Printing Process

In the FDM 3D printing process, the printer first deposits to continuous perimeter outlines of the layer being printed. It then deposits the inside of the layer based on infill setting. As can be seen Fig. 13.1, the infill value determines how the printer prints the internal structure of the piece. In SLA printing, a UV laser cures a layer of liquid resin which is force adhered to the previous layer. The resulting piece is solid, only resolution (layer thickness) can be altered. Table 13.1 shows all specimen types that were analyzed in the experiment.

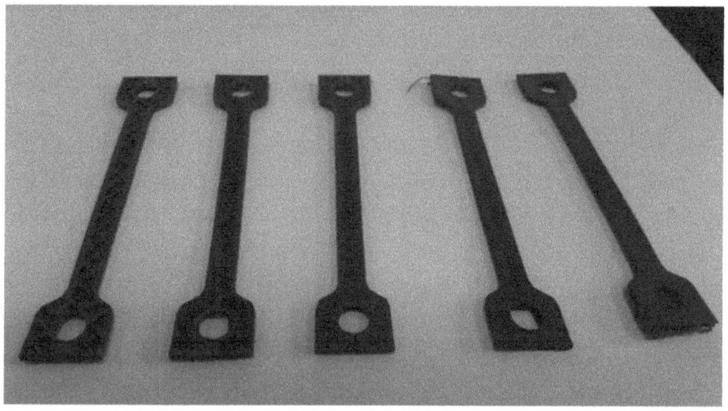

Fig. 13.1 20, 40, 60, 80, 100% infill from left to right

Table 13.1 Specimen material, infill, and print orientation types

Specimen Types								
Material / Infill	**PLA (0/90°)**	**PLA (45/-45°)**	**Tough PLA (0/90°)**	**Tough PLA (45/-45)**	**ABS (0/90°)**	**ABS (45/-45°)**	**Black Resin (Cured)**	**Black Resin (Uncured)**
	20%	20%	20%	20%	Solid	Solid	0.1	0.1
	40%	40%	40%	40%	Sparse Double	Sparse Double	0.025	0.025
	60%	60%	60%	60%	Sparse High	Sparse High		
	80%	80%	80%	80%	Sparse Low	Sparse Low		
	100%	100%	100%	100%				

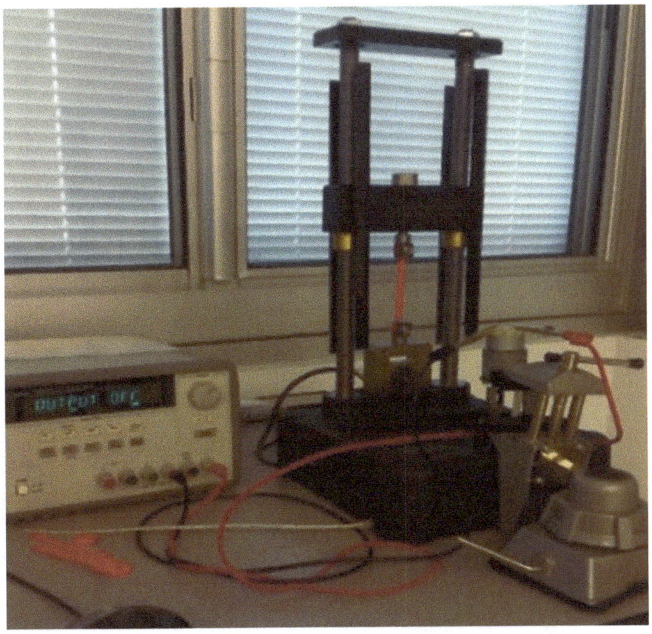

Fig. 13.2 Testing setup

13.2.3 Experimental Setup

The tensile testing machine, Pasco ME-8236 Materials Testing Apparatus, as can be seen in Fig. 13.2, was modified to use a DC 12V 3.5RPM High Torque Motor to actuate the tensile machine. The motor was calculated to be able to turn the manual hand crank at the ASTME Standard speed of 2 millimeters per minute. The substitution of the motor also eliminated any possible error that may result from using the manual hand crank method. The tensile testing machine utilized the software, Pasco Capstone, to record data at a frequency of 5 Hz. In recording data at this frequency, a data set of 500–700 points were recorded. The specimens were printed in sets of four allowing for multiple tests of each sample type [2]. This eliminated major possibilities of error in the sample sets that may appear during tensile tests [1].

13.3 Results

In calculating the stress experienced by the specimen to plot the σ-ε curves, the actual cross section of the specimen was considered. Therefore the 20% infill had a smaller cross-section than the 100% and so on, but the outer width and thickness were the same.

13.3.1 PLA Specimen

Figure 13.3 shows tensile behavior of PLA at a 0/90 configuration with various percent infills. The 20 and 100% infill behavior is similar to each other. This can be attributed to the shelling process of the 3d printer used which prints with two continuous outer shells as well as can be seen in Fig. 13.1, the 20% infill is produced in a diagonal cross pattern whereas the 40, 60, and 80% are produced in length wise grid pattern.

Figure 13.4 displays tensile behavior of PLA in a −45/45 configuration for the listed infill values. The 20, 40, 60, and 80% infill had very similar material behavior. This may be attributed to the change in printing direction.

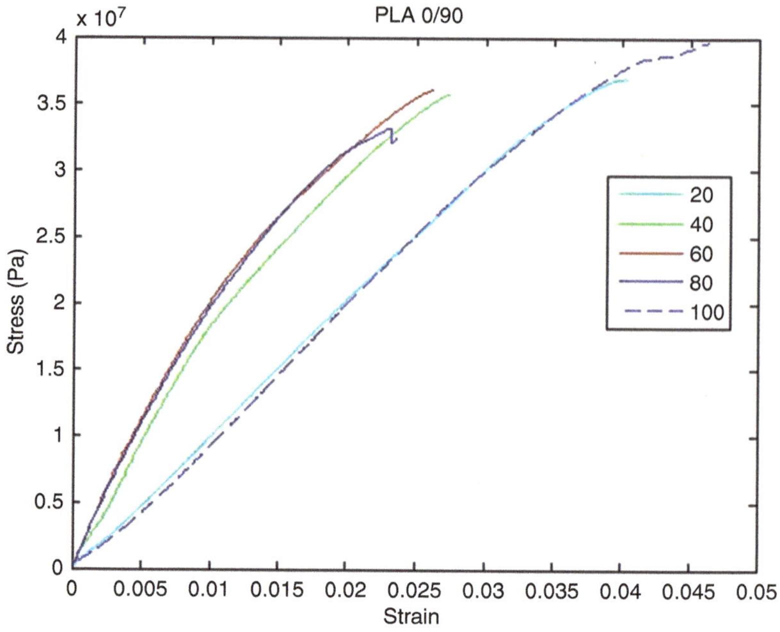

Fig. 13.3 PLA 0/90° specimen stress (σ) vs. strain (ε) graph

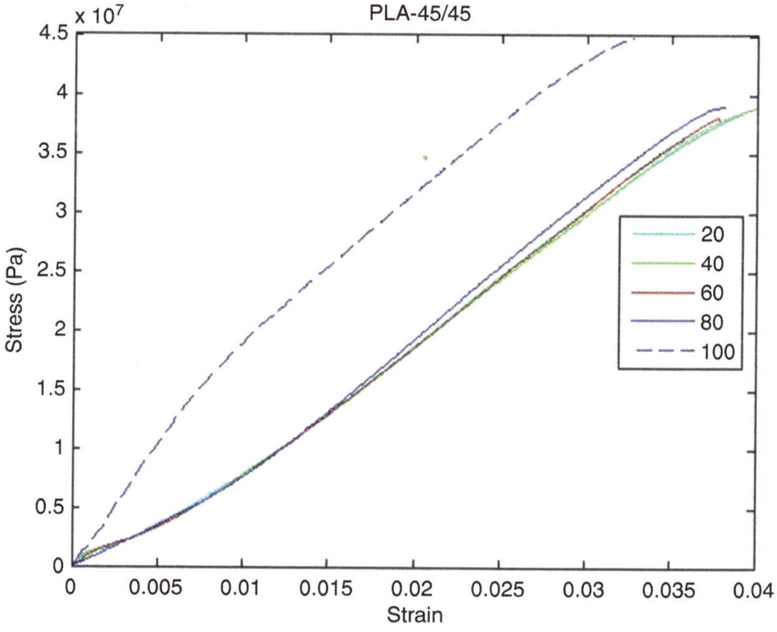

Fig. 13.4 PLA −45/45° specimen stress (σ) vs. strain (ε) graph

13.3.2 Tough PLA Specimen

Figure 13.5 shows the longitudinal stress-strain curves for the tough PLA material at the five set infill percentages. The first portion of the test showed that all five tough PLA specimen types exhibited similar behavior under tensile load. As expected, the 100% solid infill specimen had the highest ultimate strength compared to those with a lower infill.

Figure 13.6 shows tough PLA material printed at the −45/45 orientation. The resulting data showed that the 45° configuration increased the materials ultimate strength. In comparison between the graphs of Figs. 13.3 and 13.4, it was seen that the angle change in printing resulted in higher ultimate stress and yield point. The ductility increased with greater infill, unlike the 0/90 configuration which remained similar across most of the infill values.

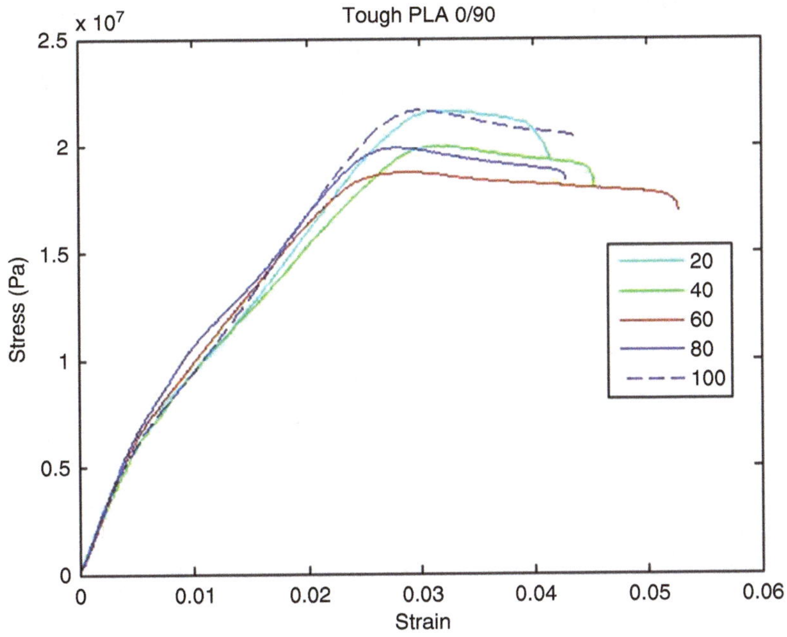

Fig. 13.5 Tough PLA 0/90° specimen stress (σ) vs. strain (ε) graph

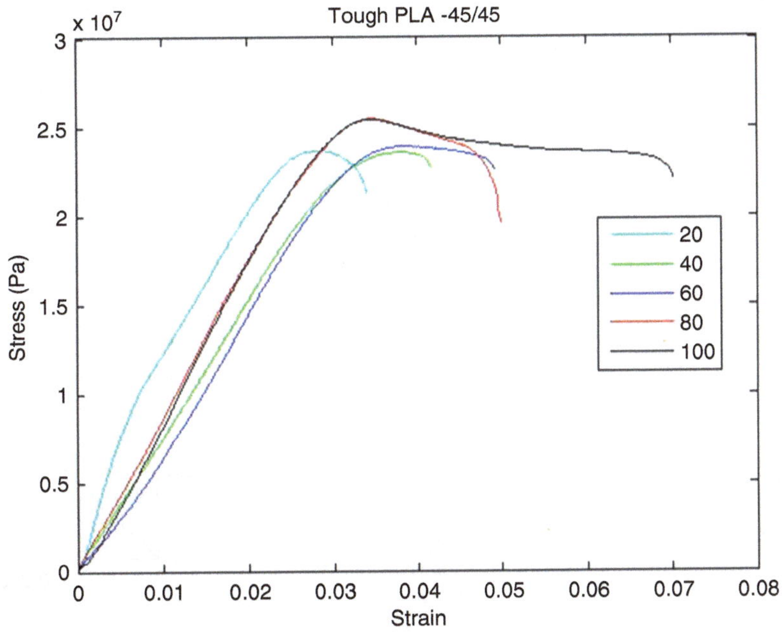

Fig. 13.6 Tough PLA −45/45° specimen stress (σ) vs. strain (ε) graph

13.3.3 ABS

The ABS FDM printer provides option of solid (SO), High density (HD), Low density (LD), and Sparse double density (DD) which follow a decreasing infill in the order listed. It can be seen from Fig. 13.5 that as the infill decreases the ultimate strength decreases and produces a reduction of the elastic modulus (Fig. 13.7).

As can be seen in Fig. 13.8, the −45/45 orientation produces a more consistent behavior as compared to the 0/90 configuration.

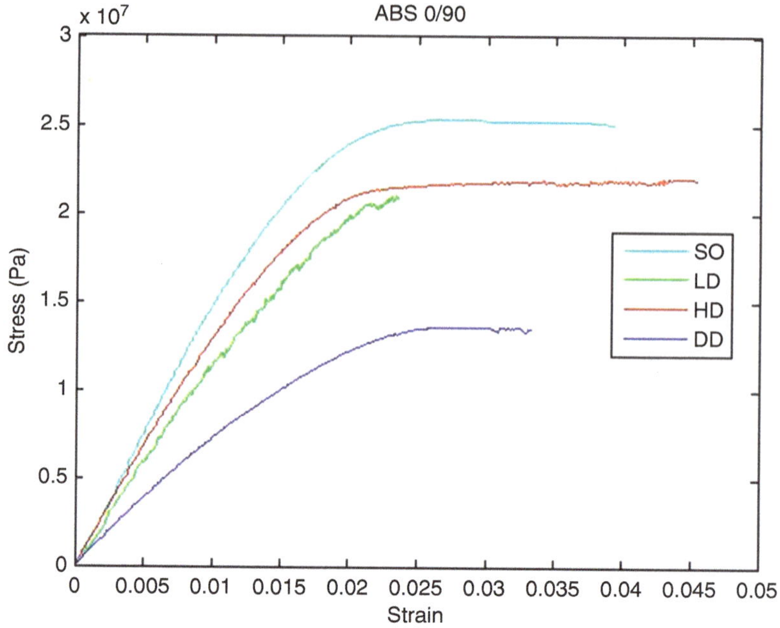

Fig. 13.7 ABS 0/90° specimen stress (σ) vs. strain (ε) graph

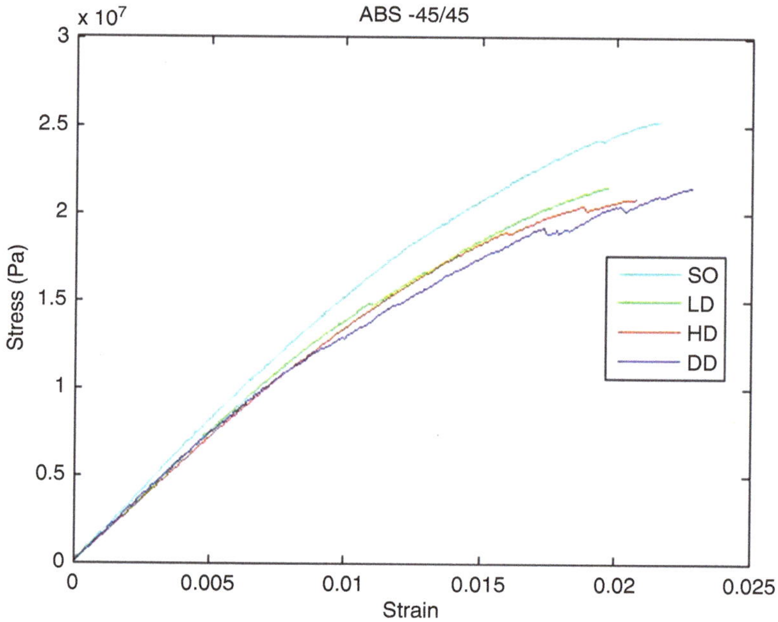

Fig. 13.8 ABS −45/45° specimen stress (σ) vs. strain (ε) graph

13.3.4 Resin

In Fig. 13.9, high resolution printing resulted in a higher ultimate strength but lower ductility. This may be attributed to the layer height of the print which is 0.1 mm for low resolution, and 0.025 mm for high resolution. This means there are four times as many layer intersection and therefore four times as many weaker layers of failure.

Again, in Fig. 13.10, the high-resolution print resulted in a higher ultimate strength, but the act of curing appears to stabilize the layers and even the ductility. The low resolution became more brittle while the high resolution became more ductile.

Table 13.2 shows a summary of all the results obtained from the testing of specimens produced by additive manufacturing.

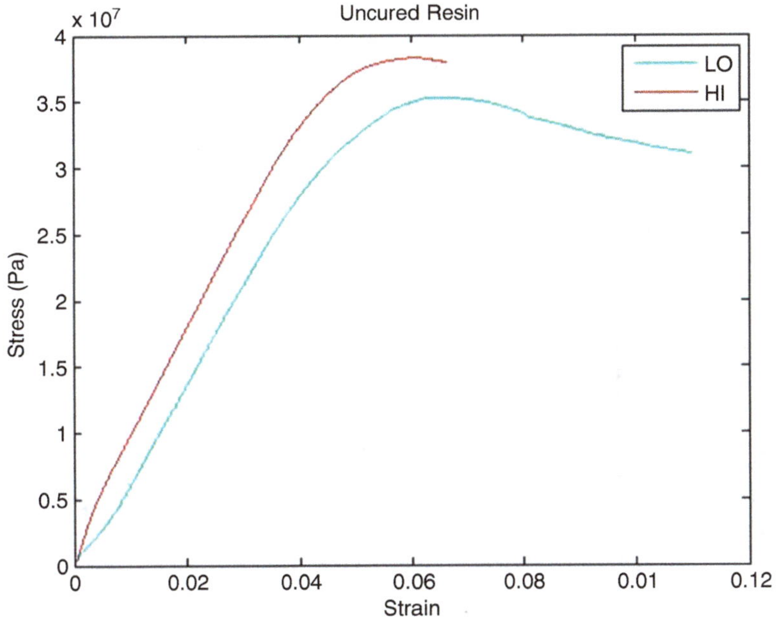

Fig. 13.9 Uncured black resin stress (σ) vs. strain (ε) graph

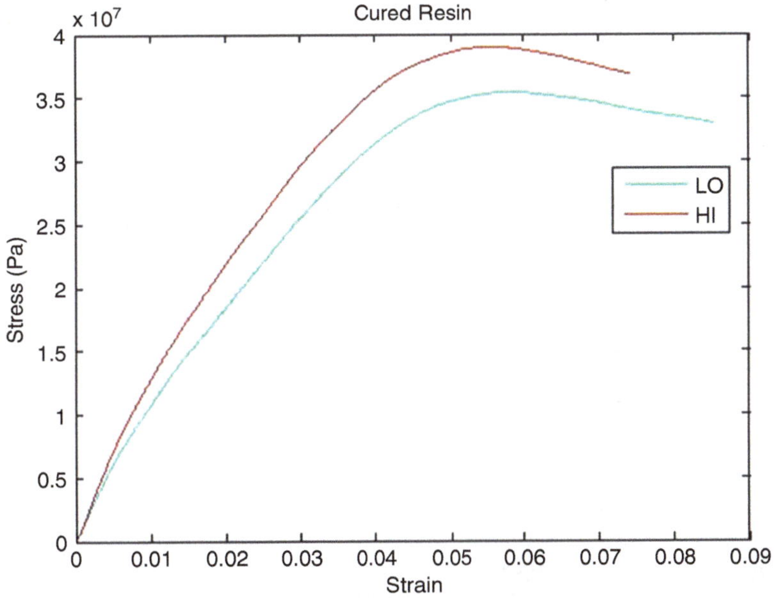

Fig. 13.10 Cured black resin stress (σ) vs. strain (ε) graph

13.4 Conclusion

The tensile testing shows that the infill and orientation of 3D printed materials effects the tensile properties of the material. The ±45° samples performed with an overall higher ultimate strength in both PLA and Tough PLA. The orientation had little effect on the ABS. This may be attributed to the grade of printer used i.e. consumer grade vs. industrial grade. In resins, a higher resolution equated to a higher ultimate stress and curing of the resin normalized the ductility of the material.

Table 13.2 Tensile properties of each material configuration

Tensile Properties of Each Material Configuration					
		Elastic Modulus (GPa)	**Sigma Max (MPa)**	**Yield Point (MPa)**	**Ductility**
PLA 0/90°	**20%**	0.9095	36.971		4.03%
	40%	1.8573	35.745		2.74%
	60%	2.2815	36.089	22.81	2.61%
	80%	2.207	33.2	21.54	2.35%
	100%	0.8004	39.673		4.63%
PLA - 45/45°	**20%**	0.6475	38.349		3.91%
	40%	0.5812	38.845		4.00%
	60%	0.6205	38.093		3.78%
	80%	0.6926	39.026		3.82%
	100%	2.1316	44.836	22.52	3.27%
Tough PLA 0/90°	**20%**	1.2648	21.619	8.734	4.16%
	40%	1.1739	19.965	9.289	4.54%
	60%	1.2767	18.763	9.421	5.28%
	80%	1.3224	19.901	10.65	4.29%
	100%	1.2333	21.64	8.325	4.41%
Tough PLA -45/45°	**20%**	1.5632	23.638	10.201	3.40%
	40%	0.7233	23.539	23.01	4.16%
	60%	0.5584	23.892	23.63	4.91%
	80%	0.8185	25.46	25.23	5.00%
	100%	0.8332	25.414	25.29	7.02%
ABS 0/90°	**Solid**	1.5132	25.301	21.5	3.93%
	Low Density	1.212	20.972	16.99	2.36%
	High Density	1.3688	21.959	17.85	4.54%
	Double Density	0.779	13.57	9.766	3.33%
ABS -45/45°	**Solid**	1.6317	25.276	19.99	2.17%
	Low Density	1.4753	21.486	19.18	1.97%
	High Density	1.4243	20.796	18.55	2.07%
	Double Density	1.4797	21.438	14.57	2.28%
Uncured Resin	**Low Resolution**	0.5662	35.338		11%
	High Resolution	1.1965	38.295	10.57	6.59%
Cured Resin	**Low Resolution**	1.2334	35.394	12.91	8.53%
	High Resolution	1.4595	38.967	15.39	7.42%

Acknowledgements The authors would like to thank the President of Vaughn College of Aeronautics and Technology, Dr. Sharon DeVivo and the Department Chair of Engineering and Technology, Dr. Hossein Rahemi, for their enthusiastic support. Furthermore, the authors would like to thank Professor Manuel Jesus for support and assistance in the operation and use of the 3D printers and his support in this research. The authors would also like to thank adjunct professor Christopher D. Griffith for his assistance and support in this research.

References

1. Budhoo, Y., Delale, F., Liaw, B.: Temperature effect on tensile testing of hybrid and non-hybrid composites. To be published (SEM 2011 Annual Conference & Exposition on Experimental and Applied Mechanics)
2. Budhoo, Y., Delale, F., Liaw, B.: Effect of temperature on the Poisson's ratio of woven composites. To be published (SEM 2011 Annual Conference & Exposition on Experimental and Applied Mechanics)

Chapter 14
Tool Wear Mechanisms of Physical Vapor Deposition (PVD) TiAlN Coated Tools Under Vegetable Oil Based Lubrication

Salman Pervaiz and Wael Abdel Samad

Abstract Titanium alloys are mainly utilized in the aerospace, biomedical, power generation, automotive and marine sectors due to their high hot hardness, ability to operate at higher operating temperatures and resistance against corrosion. Alpha-beta titanium alloy (Ti6Al4V) is the most frequently used titanium alloy, mainly preferred for the applications such as blades, discs, jet engine airframes and biomedical implants. TiAlN coating is highly favored for the metal cutting applications due to its superior cutting performance. The high cutting performance of TiAlN coating is linked with the addition of aluminum to the traditional TiN coating. The addition of aluminum results in the formation of aluminum oxide on the cutting tool surface resulting in ability to sustain high cutting temperatures. In this study a series of machining experiments were conducted to evaluate the machinability of difficult-to-cut titanium alloy (Ti6Al4V) by using PVD-TiAlN coated carbide tools under vegetable oil assisted minimum quantity cooling lubrication (MQCL). The cutting insert consists of substrate based on tungsten carbide with 6% cobalt to maintain high hot hardness and fracture resistance. Thin PVD-TiAlN coating on the substrate provides excellent adhesion resistance to prolong the cutting edge sharpness and results in enhanced tool life. In this study the influences of vegetable oil flow rate was taken into consideration to evaluate machinability. The study incorporated two different oil flow rates of 60 and 80 ml/h using a specially designed tool holder with internal coolant delivery channels for MQCL arrangement. The MQCL based results were also compared with the dry cutting. Detailed study on micro-wear mechanisms at the cutting edge was conducted by utilizing scanning electron microscopy (SEM). The study revealed adhesion, micro-edge chipping, abrasion and attrition as major wear mechanisms when using PVD-TiAlN coated tools for machining Ti6Al4V.

Keywords MQCL • TiAlN coating • Ti6Al4V • Vegetable oil • Wear mechanisms

14.1 Introduction

Titanium alloys are termed as one of the high performance alloys due to their low density, high strength to weight ratio, extraordinary fracture strength and superior corrosion resistance. Due to these characteristics titanium alloys are mainly employed in aerospace, energy, automotive, biomedical, transportation and marine applications. At the same time, they represents poor machinability due to some of their inherent properties such as low thermal conductivity, high hot hardness and chemical reactivity towards all cutting tool materials [1–4]. Nowadays in the metal cutting industry, most of the research activities have been dedicated towards attaining improved machining performance through higher material removal rate (MMR), unconventional cutting tool materials, hybrid cutting environments and surface integrity. However, these improvements should also incorporate environmental and health based sustainability aspects to achieve cleaner and benign outcomes. Several international platforms such as EU, OECD and UN are taking well directed initiatives to facilitate sustainable development. Implementation of sustainability concepts in the metal cutting sector also improves the overall economics of the process as well. These sustainability concepts are incorporated in the metal cutting sector by executing the green practices such as decreasing the input resources (energy, material and water etc.), minimizing the use of toxic and non-biodegradable substances and providing better ergonomics, health/ safety facilities, and employee personal development [5]. The sustainable machining concepts also aim to minimize the amount of greenhouse gas (GHG) emissions and related carbon footprints. At the same time, toxic and non-biodegradable nature of the cutting fluid is also being condemned. When it comes to the sustainable machining of titanium alloys, most of the studies were executed to reduce the energy consumption involved and possibilities to replace conventional flood cooling method with alternate environment friendly cooling/ lubrication approaches.

S. Pervaiz • W.A. Samad (✉)
Department of Mechanical and Industrial Engineering, Rochester Institute of Technology – Dubai, 341055 Dubai, UAE
e-mail: wascad@rit.edu

© The Society for Experimental Mechanics, Inc. 2018

J. Wang et al. (eds.), *Mechanics of Additive and Advanced Manufacturing, Volume 9*,
Conference Proceedings of the Society for Experimental Mechanics Series, DOI 10.1007/978-3-319-62834-9_14

As reported [6, 7] the global consumption of metal working fluids (MWFs) is 640 million gallons per annum, where 52% is being used in the metal cutting sector only. When machining titanium alloys, application of cutting fluid is considered essential to maximize the tool life and improve the surface integrity. At the same time cutting fluids help in the removal of chips. As per the strict environmental laws and regulations, the usage of the cutting fluid is highly questioned and discouraged because most of these cutting fluids pose danger to the health and environment. To achieve sustainability in the machining process, it is essential to use non-toxic and biodegradable cutting fluids. The below section reviews the potential of different sustainable cooling/ lubrication strategies when machining titanium alloys. The concepts of minimum quantity lubrication (MQL) and minimum quantity cooling lubrication (MQCL) were proposed in the metal cutting sector almost a decade back, the main idea was to cater the negative influence of conventional flood cooling method. Several researchers have explored the potential of MQL for the metal cutting sector and found very encouraging results. Syed et al. [8] studied the machining performance of Ti6Al4V using six different cooling and lubrication techniques. The study was performed using dry, MQL, MQCL (internal and external), cryogenic and cooled air strategies. The study exposed promising potential MQL and MQCL to replace conventional flood cooling method.

Pejryd et al. [9] studied the machinability of high performance aeronautic alloys, including Ti6Al4V, by consuming ester and fatty alcohol based MQCL arrangements with uncoated and ceramic based cutting tools. The study exposed improved tool life when machining Ti6Al4V using MQCL technique. Ramana et al. [10] also assessed machinability of titanium alloy under MQL setup. The study utilized a tool wear optimization approach based on Taguchi's robust design method. The study found that cutting speed has controlling influence on tool life. Pervaiz et al. [11] investigated the machining performance of Ti6Al4V under emulsion based MQL setup. The study was conducted using coated and uncoated cutting tools. The study revealed encouraging potential at higher cutting speeds. Liu et al. [12] utilized nano composite (nc-AlTiN)/ (a-Si3N4) and (nc-AlCrN)/(a-Si3N4) coated cutting tool when machining Ti6Al4V. The study observed that (nc-AlTiN)/(a-Si3N4) coating performed better than (ncAlCrN)/(a-Si3N4) coating. Wang et al. [13] studied the machinability of Ti6Al4V using dry, MQL and flood cooling methods in continuous and interrupted machining operations. The study estimated friction coefficient at tool chip interface using the cutting forces. It was observed that MQL strategy performed better at higher cutting speeds in case of interrupted machining. Pervaiz et al. [14] in another study utilized vegetable oil based minimum quantity cooling lubrication (MQCL) setup to investigate the machinability of Ti6Al4V. The work was carried to study both external and internal arrangements of MQCL. The study revealed promising machining performance of MQCL (internal) when compared with conventional flood cooling environments.

Park et al. [15] investigated the machinability of Ti6Al4V by using MQL and cryogenic cooling methods. Besides the conventional MQL cooling, the study also utilized a novel approach where MQL was mixed with exfoliated graphite nano-platelets (xGnPs). The machining experiments were conducted using the face and end milling arrangements. The results obtained in cryogenic and MQL cooling methods were compared with the conventional dry and wet cutting as well. The cryogenic method was more oriented towards providing the cooling action during the machining of Ti6Al4V. However, the MQL method was mainly providing the lubrication during the cutting process. The study revealed that tool wear and cutting forces were low in the start for cryogenic method. With the passage of time when workpiece get hard both tool wear and cutting forces increased rapid due to adhesion and chipping mechanisms. The study also revealed that MQL combined with graphite nano-platelets (xGnPs) outperformed other methods. The reason can be linked with improved lubrication provided by graphite nano-platelets (xGnPs). Kim et al. [16] performed micro-end milling process on Ti6Al4V using nanofluid MQL and CO_2 chilly gas cooling methods. The study was focused to measure cutting forces, surface finish, tool wear and coefficient of friction. The study revealed that low concentration (0.1%) of nanoparticles based MQL provided low cutting forces and tool wear, whereas high concentration (1.0%) of nanoparticles based MQL provided high surface finish. Mathew and Vijayaraghavan [17] conducted MQL assisted drilling experiments on intermetallic titanium aluminide alloy. The study considered high and low aspect ratios. The study considered thrust force and torque as the main output parameter. Chip morphology is also very important as it has significant influence on the thrust force and associated torque. The study also complimented on the burr formation under MQL. The MQL results were compared with dry and wet cooling methods. Gupta et al. [18] conducted MQL based study to machine aeronautic titanium alloy (grade 2). The study combined experimental observations with empirical modeling technique to draw conclusions. In addition the study complimented the output results using desirability optimization technique. The study utilized response surface methodology (RSM) in the first phase to perform machining experiments using different levels of cutting parameters and approach angles. In the second phase analysis of variance (ANOVA) and regression analysis were conducted. The study found optimized cutting condition to be at 200 m/min, 0.10 mm/ rev feed rate and 90° cutting edge angle.

The current study is focused on the machinability investigation of Ti6Al4V using PVD-TiAlN coated carbide tools. The study utilized advanced MQL assisted concept where conventional MQL was combined with the chilled air (−4 C°). The approached is termed as minimum quantity cooling lubrication (MQCL). The study incorporated two different oil flow rates of 60 and 80 ml/h using a specially designed tool holder with internal coolant delivery channels for MQCL arrangement. The resulted outcomes of MQCL machining were compared against the conventional dry cutting and flood cooling methods.

14.2 Experimental Setup and Cutting Conditions

The cutting experiments were conducted on the most commonly used titanium alloy (Ti6Al4V) using turning arrangement. The raw material of Ti6Al4V was available in the form of cylindrical rods with ASTM B381standard specifications. It is also important to understand the importance of Ti6Al4V because almost 50% of the total titanium consumption is based on Ti6Al4V alloy. Table 14.1 reports the chemical composition and mechanical properties of the raw workpiece material.

The machining experiments were executed using the PVD-TiAlN coated cutting tools. TiAlN coating is favored when machining titanium alloys due to its resistance against the tool wear. To support the idea of internal coolant MQCL delivery, a specially designed tool holder has been utilized from Mircona. The specifications of the cutting tool, tool holder and vegetable oil for MQL system have been reported in Table 14.2.

Table 14.1 Specifications of workpiece material (Ti6Al4V)

(A) Chemical composition							
Element	Wt. %	Element	Wt. %	Element	Wt. %	Element	Wt. %
H	0.015	V	3.5–4.5	C	0.1	Ti	Balance
N	0.05	Al	5.5–6.18	Fe	0.3		

(B) Mechanical properties of Ti6Al4V at room temperature			
Properties	Values	Properties	Values
Ultimate tensile strength	993 MPa	Poison ratio	0.342
Yield point	830 MPa	Young's modulus	114 GPa
Elongation	14	Hardness (HRC)	36

Table 14.2 Cutting tool and tool holder specifications

(A) Cutting tool specifications		(B) Cutting parameters	
Make	Sandvik	Cutting speed	90 m/min
ISO code	CCMT 12 04 04 mm 1105	Feed	0.1 mm/rev
Material	Cermet inserts	Depth of cut	0.8 mm
Rake	Positive	Cooling strategies	Dry, flood & MQCL air at (−4 C°) was mixed
Coating	PVD	MQCL – oil flow rates	60, 80 ml/h

(C) Tool holder and vegetable oil for MQCL	
Make	*Mircona*
Tool holder model	SCLC R 2525 M12-EB
Vegetable oil supplier	*ACCU-Svenska AB*
Type	Rape seed oil ECULUBRIC E200L
Chemical description	Triglycerides, easily biodegradable substances
Health hazard	Not hazard to human health
Flash point	325 C°
Ignition point	365 C°
Density (at 0 C°)	0.9273 g/cm^3
(At −4 C°)	0.9297 g/cm^3

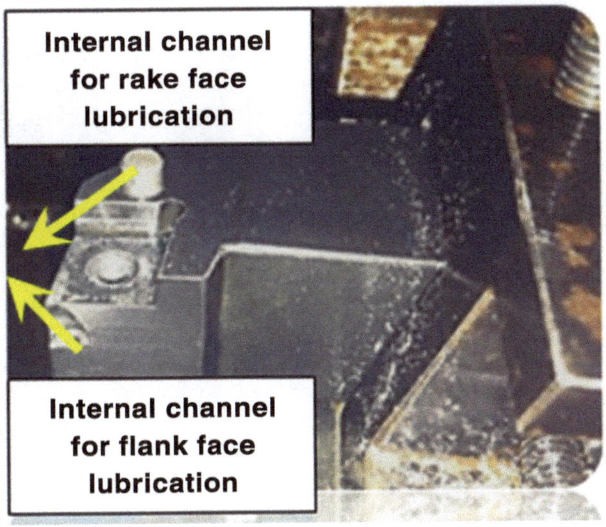

14.3 Results and Discussion

14.3.1 *Flank Tool Wear and Associated Tool Life*

In start of the machining process, the cutting edge of the tool is very sharp in nature that facilitates the cutting operation. But with the passage of time the sharp cutting edge becomes dull and it is linked with the initiation of different types of wear mechanisms. In metal cutting industry, the wear on the flank face of the cutting tool is easier and most commonly used criterion to evaluate the tool life. Flank tool wear is mainly produced due to the rubbing action the machine surface and flank face of the cutting tool. At the same time flank tool wear plays a significant role towards the overall performance of the cutting process. Generally cutting forces are linked with the flank tool wear progression. Increase in the flank wear results in increasing the cutting force as well. Once the critical value of flank tool wear is reached the higher magnitude of cutting force can break the cutting tool as well (Fig. 14.1).

The cutting experiments analyzed flank tool wear as elaborated by the ISO3685 standard [19] using the tool maker microscope. Flank tool wear curves for dry, flood, MQCL-60 ml/h and MQCL-80 ml/ h have been plotted as shown in the Fig. 14.2. It can be observed in Fig. 14.2 that the critical value of flank tool wear was taken at 0.3 mm. It can be observed that the flank tool wear curve obtained for the dry cutting provided the shortest cutting tool life. The short tool life in dry cutting can be explained by keeping in view the aggressive nature of contact present between the cutting tool and workpiece material without any coolant/ lubricant. The presence of high friction during the dry cutting results in high cutting temperature at the cutting interface. Higher cutting temperature can reduce the tool life by initializing different tool wear mechanism such as adhesion, diffusion and plastic softening of the sharp cutting edge etc. The tool wear examination will provide more understanding in this case of dry cutting. As expected flood cooling method provided the highest tool life. The better performance in this case can be linked with the efficient cutting temperature control at the cutting interface. The regular heat dissipation using flood coolant enhanced the tool life and flak wear progression was gradual. It can be observed that MQCL-60 ml /h provided more or less the similar tool life as observed during the flood cooling method. It means that that MQCL assisted cooling method has a good potential to replace existing flood cooling. This advancement will be helpful in making the metal cutting process green in nature. The performance of MQCL-80 ml/h represents that there is a concept of optimized flow rate when it comes to the MQCL cooling. It is due to the reason that oil flow rate in MQCL is linked with the parameters such as penetration ability and the lubrication capacity. Better performance can only be achieved at certain level of agreement between these two parameters.

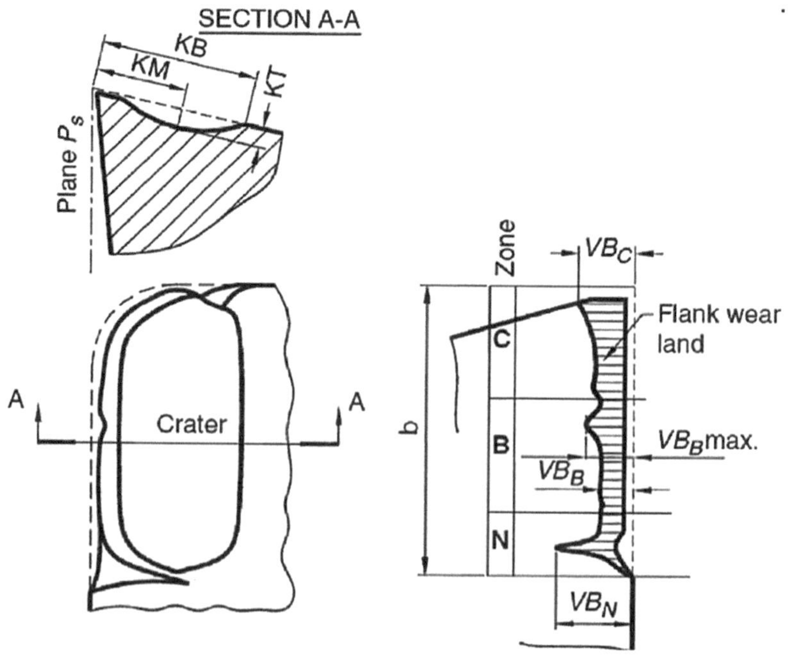

Fig. 14.1 Schematic illustration of tool wear patterns (adopted from ISO Standard 3685) [19]

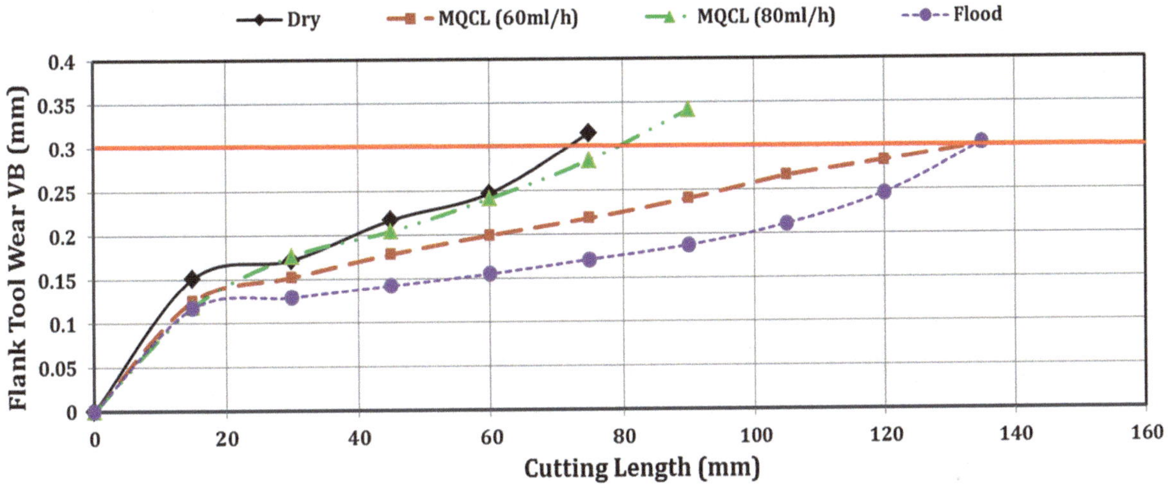

Fig. 14.2 Flank tool wear curves for the dry, flood, MQCL-60 ml/h and MQCL-80 ml/h

14.3.2 Tool Wear Mechanisms

In order to investigate the tool wear mechanisms associated with the TiAlN coated tool, scanning electron microscopy has been utilized as shown in the Fig. 14.3. Figure 14.3a represents presence of adhesion as a major mechanism but at the same time edge chipping can also be observed. Slight traces of abrasion wear can also be observed. The wear profile in Fig. 14.3a under dry cutting shows the presence of high cutting temperature. Figure 14.3b represents the micrograph of MQCL-80 ml/h, it can be observed that in this case a large portion of cutting edge has been chipped off. This represents the presence of high contact pressure at the cutting edge. At the same time comparatively high adhesion has been observed. The higher adhesion represents the presence of high cutting temperature at the cutting interface. Figure 14.3b also reports that it is not necessary that higher level of oil flow rate (80 ml/ h) under MQCL arrangement always improves the machining performance. This means that higher level of oil flow rate can decrease the penetration ability of the spray as a result the cutting interface is not properly lubricated. In Fig. 14.3c, the tool wear for MQCL – 60 ml/h has been reported. It can be observed that adhesion in this case, is comparatively less than the MQCL – 80 ml/ h case. It represents that spray was able to dissipate heat properly at 60 ml/h that can be linked with improved penetration ability of the spray. Similar trend has been observed for the cutting edge chipping for MQCL – 60 ml/ h. It is due to the reason that cutting edge has catered less contact pressure or friction control was more efficient in this case. Figure 14.3d reports the tool wear mechanisms for the traditional flood cooling method. As expected flood cooling method outclassed all other cooling methods. However the performance of MQCL – 60 ml/h was found very close to the flood cooling method.

14.4 Conclusions

As mentioned below, following conclusions have been drawn from the current study.

- The dry cutting environment provided the lowest tool life among other cooling methods and adhesion and chipping were found as major tool wear mechanisms.
- The study compared the machining performance of Ti6Al4V under MQCL arrangement with two flowrate of 60 and 80 ml/h. The optimum performance of MQCL arrangement was found with the low oil flow rate of 60 ml/ h. It reveals that penetration ability and lubrication capacity parameters play importance role towards oil flow rate optimization.
- The experimental combination of PVD-TiAlN based coating and MQCL cooling method has a potential to replace the flood cooling method. This combination can be further explored to make machining of Ti6Al4V more sustainable in nature.

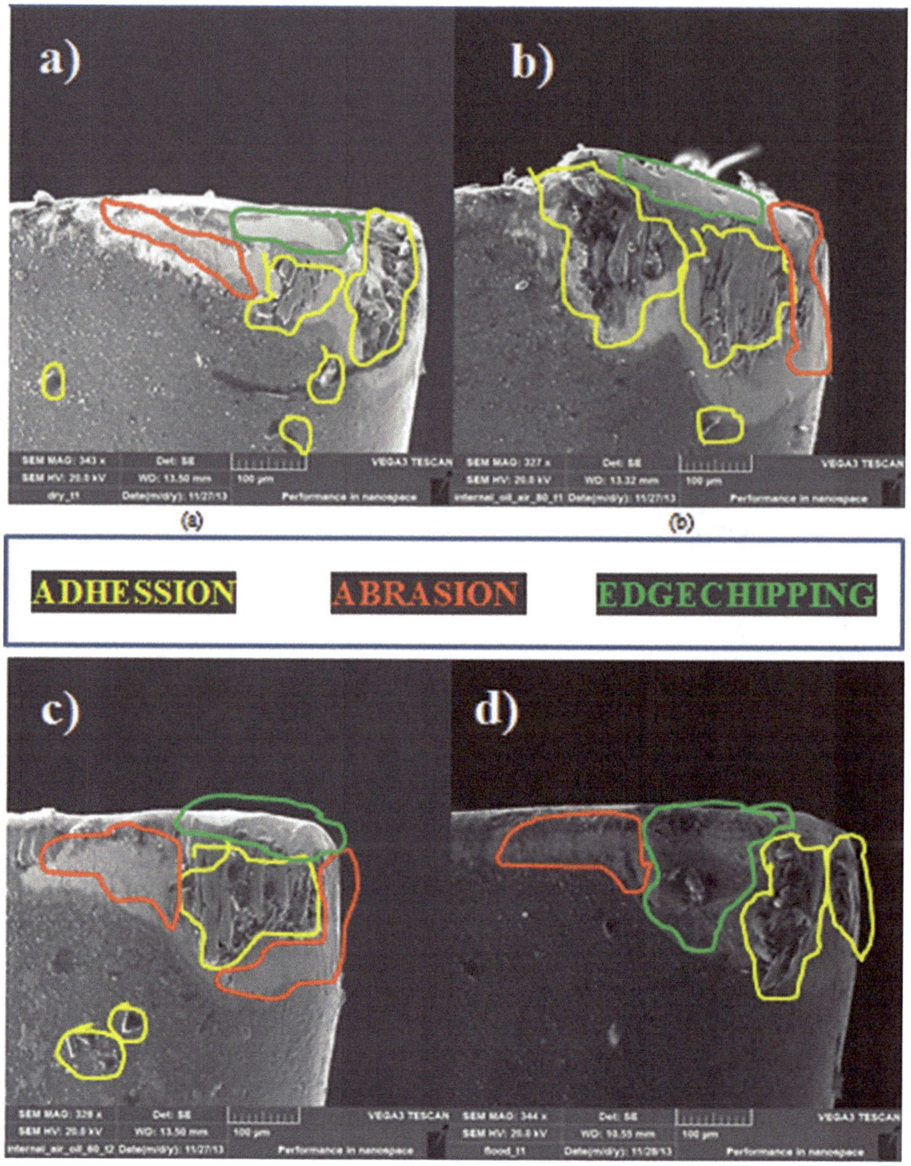

Fig. 14.3 Scanning electron micrographs of PVD-TiAlN coated carbide cutting tool (**a**) dry, (**b**) MQCL-80 ml/h, (**c**) MQCL-60 ml/h and (**d**) flood cooling

References

1. Ezugwu, E.O., Wang, Z.M., Machado, A.R.: The machinability of nickel-based alloys: a review. J. Mater. Process. Technol. **86**, 1–16 (1999)
2. Ezugwu, E., Bonney, J., Yamane, Y.: An overview of the machinability of aeroengine alloys. J. Mater. Process. Technol. **134**(2), 233–253 (2003)
3. Ezugwu, E.O., Wang, Z.M.: Materials titanium alloys and their machinability. J. Mater. Process. Technol. **68**, 262–274 (1997)
4. Nabhani, F.: Machining of aerospace titanium alloys. Robot. Comput. Integr. Manuf. **17**, 99–106 (2001)
5. Jayal, A.D., Badurdeen, F., Dillon, O.W., Jawahir, I.S.: Sustainable manufacturing: modeling and optimization challenges at the product, process and system levels. CIRP J. Manuf. Sci. Technol. **2**(3), 144–152 (2010)
6. Marksberry, P.W., Jawahir, I.S.: A comprehensive tool-wear/tool-life performance model in the evaluation of NDM (near dry machining) for sustainable manufacturing. Int. J. Mach. Tools Manuf. **48**(7–8), 878–886 (2008)
7. Brockhoff, T., Walter, A.: Fluid minimization in cutting and grinding. Abrasives Mag. **Oct/Nov**, 38–42 (1998)
8. Raza, S.W., Pervaiz, S., Deiab, I.: Tool wear patterns when turning of titanium alloy using sustainable lubrication strategies. Int. J. Precis. Eng. Manuf. **15**(9), 1979–1985 (2014)
9. Pejryd, L., Beno, T., Isaksson, M.: Machining aerospace materials with room-temperature and cooled minimal-quantity cutting fluids. Part B J. Eng. Manuf. **225**, 74–86 (2010)

10. Venkata, R.M., Krishna Mohan, R.G., Hanumantha, R.D.: Optimization and effect of process parameters on tool wear in turning of titanium alloy under different machining conditions. Int. J. Mater. Mech. Manuf. **2**(4), 272–277 (2014)
11. Pervaiz, S., Deiab, I., Darras, B.: Power consumption and tool wear assessment when machining titanium alloys. Int. J. Precis. Eng. Manuf. **14** (6), 925–936 (2013)
12. Liu, Z., An, Q., Xu, J., Chen, M., Han, S.: Wear performance of (nc-AlTiN)/(a-Si3N4) coating and (nc-AlCrN)/(a-Si3N4) coating in high-speed machining of titanium alloys under dry and minimum quantity lubrication (MQL) conditions. Wear. **305**(1–2), 249–259 (2013)
13. Wang, Z.G., Rahman, M., Wong, Y.S., Neo, K.S., Sun, J., Tan, C.H., Onozuka, H.: Study on orthogonal turning of titanium alloys with different coolant supply strategies. Int. J. Adv. Manuf. Technol. **42**(7–8), 621–632 (2009)
14. Pervaiz, S., Deiab, I., Rashid, A., Nicolescu, M.: Minimal quantity cooling lubrication in turning of Ti6Al4V: influence on surface roughness, cutting force and tool wear. Proc. Inst. Mech. Eng. Part B J. Eng. Manuf. **231**(9), (2017)
15. Park, K., Suhaimi, M.A., Yang, G., Lee, D., Lee, S., Kwon, P.: Milling of titanium alloy with cryogenic cooling and minimum quantity lubrication (MQL). Int. J. Precis. Eng. Manuf. **18**(1), 5–14 (2017)
16. Kim, J.S., Kim, J.W., Lee, S.W.: Experimental characterization on micro-end milling of titanium alloy using nanofluid minimum quantity lubrication with chilly gas. Int. J. Adv. Manuf. Tech. **91**, 2741–2749 (2017)
17. Mathew, N.T., Vijayaraghavan, L.: Environmentally friendly drilling of intermetallic titanium aluminide at different aspect ratio. J. Clean. Prod. **141**, 439–452 (2017)
18. Kumar, M., Kumar, P., Singh, G., Sharma, V.S.: Sustainable machining of aerospace material e Ti (grade-2) alloy: modeling and optimization. J. Clean. Prod. **147**, 614–627 (2017)
19. ISO 3685, Tool life testing with single-point turning tools, ISO Standard, 3685 (1993) (E)

FSC
www.fsc.org
MIX
Papier | Fördert gute Waldnutzung
FSC® C083411

Kolibri 360 GmbH
Ferdinand-Jühlke-Straße 7
99095 Erfurt, Deutschland
produktsicherheit@kolibri360.de